Civil Engineering

Civil Engineering

Naseer Khan

RANDOM PUBLICATIONS
NEW DELHI (INDIA)

Civil Engineering

ISBN 978-93-5111-431-4

Published in 2014 in India by

RANDOM PUBLICATIONS

4376-A/4B, Gali Murari Lal, Ansari Road
New Delhi-110 002
Phone : +91-11-43580356, +91-11-23289044
e-mail: randomexports@gmail.com, sales@randompublications.com, info@randompublications.com

Reprinted 2024

Type Setting by: Friends Media, Delhi-110089
Digitally Pritnted at : Replika Press Pvt. Ltd.

Preface

Civil engineering is the application of physical and scientific principles for solving the problems of society, and its history is intricately linked to advances in understanding of physics and mathematics throughout history. Because civil engineering is a wide ranging profession, including several separate specialized sub-disciplines, its history is linked to knowledge of structures, materials science, geography, geology, soils, hydrology, environment, mechanics and other fields. Throughout ancient and medieval history most architectural design and construction was carried out by artisans, such as stonemasons and carpenters, rising to the role of master builder. Knowledge was retained in guilds and seldom supplanted by advances. Structures, roads and infrastructure that existed were repetitive, and increases in scale were incremental. One of the earliest examples of a scientific approach to physical and mathematical problems applicable to civil engineering is the work of Archimedes in the 3rd century BC, including Archimedes Principle, which underpins our understanding of buoyancy, and practical solutions such as Archimedes' screw. Brahmagupta, an Indian mathematician, used arithmetic in the 7th century AD, based on Hindu-Arabic numerals, for excavation (volume) computations.

Civil engineers typically possess an academic degree with a major in civil engineering. The length of study for such a degree is usually three to five years and the completed degree is usually designated as a Bachelor of Engineering, though some universities designate the degree as a Bachelor of Science. The degree generally includes units covering physics, mathematics, project management, design and specific topics in civil engineering. Initially such topics cover most, if not all, of the sub-disciplines of civil engineering. Students then choose to specialize in one or more sub-disciplines towards the end of the degree. While an Undergraduate (BEng/BSc) Degree will normally provide successful students with industry accredited qualification, some universities offer postgraduate engineering awards (MEng/MSc) which allow students to further specialize in their particular area of interest within engineering. In most countries, a Bachelor's degree in engineering represents the first step towards professional certification and the degree program itself is certified by a professional body. After completing a certified degree program the engineer must satisfy a range of requirements (including work experience and exam requirements) before being certified. Once certified, the engineer is designated the title of Professional Engineer (in the United States, Canada and South Africa), Chartered Engineer (in most Commonwealth countries), Chartered Professional Engineer (in Australia and New Zealand), or European Engineer (in much of the European Union). There are international engineering agreements between relevant professional bodies which are designed to allow engineers to practice across international borders. The advantages of certification vary depending upon location. For example, in the United States and Canada "only a licensed professional engineer may prepare, sign and seal, and submit engineering plans and drawings to a public authority for approval, or seal engineering work for public and private clients.". This requirement is enforced by state and provincial legislation such as Quebec's Engineers Act. In other countries such as the UK no such legislation exists. In Australia, state licensing of engineers is limited to the state of Queensland. Practically all certifying bodies maintain a code of ethics that they expect all members to abide by or risk expulsion. In this way, these organizations play an important role in maintaining ethical standards for the profession. Even in jurisdictions where certification has little or no legal bearing on work, engineers are subject to contract law. In cases where an engineer's work fails he or she may be subject to the tort of negligence and, in extreme cases, the charge of criminal negligence. An engineer's work must also comply with numerous other rules and regulations such as building codes and legislation pertaining to environmental law.

The book provides a clear, scientific and technical understanding of this subject. The information contained herein will equip the readers with latest and up-to-date knowledge.

—*Naseer Khan*

Contents

1

Introduction to Civil Engineering

Civil engineering is a professional engineering discipline that deals with the design, construction, and maintenance of the physical and naturally built environment, including works like roads, bridges, canals, dams, and buildings. Civil engineering is the oldest engineering discipline after military engineering, and it was defined to distinguish non-military engineering from military engineering. It is traditionally broken into several sub-disciplines including environmental engineering, geotechnical engineering, structural engineering, transportation engineering, municipal or urban engineering, water resources engineering, materials engineering, coastal engineering, surveying, and construction engineering. Civil engineering takes place on all levels: in the public sector from municipal through to national governments, and in the private sector from individual homeowners through to international companies.

History of Structural Engineering

The history of structural engineering dates back to at least 2700 BC when the step pyramid for Pharaoh Djoser was built by Imhotep, the first engineer in history known by name. Pyramids were the most common major structures built by ancient civilizations because it is a structural form which is inherently stable and can be almost infinitely scaled (as opposed to most other structural forms, which cannot be linearly increased in size in proportion to increased loads).

Another notable engineering feat from antiquity still in use today is the qanat water management system. Qanat technology developed in the time of the Medes, the predecessors of the Persian Empire (modern-day Iran which has the oldest and longest Qanat (older than

3000 years and longer than 71 km) that also spread to other cultures having had contact with the Persian.

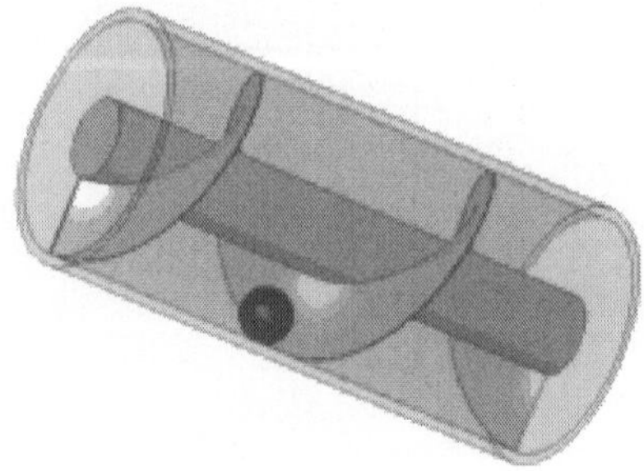

Figure: *The Archimedes screw was operated by hand and could raise water efficiently.*

Throughout ancient and medieval history most architectural design and construction was carried out by artisans, such as stone masons and carpenters, rising to the role of master builder. No theory of structures existed and understanding of how structures stood up was extremely limited, and based almost entirely on empirical evidence of 'what had worked before'. Knowledge was retained by guilds and seldom supplanted by advances. Structures were repetitive, and increases in scale were incremental.

Figure: *Pont du Gard, France, a Roman aqueduct built circa 19 BC.*

No record exists of the first calculations of the strength of structural members or the behaviour of structural material, but the profession of structural engineer only really took shape with the industrial revolution and the re-invention of concrete. The physical sciences underlying structural engineering began to be understood in the Renaissance and have been developing ever since.

Early Structural Engineering Developments

The recorded history of structural engineering starts with the ancient Egyptians. In the 27th century BC, Imhotep was the first structural engineer known by name and constructed the first known step pyramid in Egypt. In the 26th century BC, the Great Pyramid of Giza was constructed in Egypt. It remained the largest man-made structure for millennia and was considered an unsurpassed feat in architecture until the 19th century AD.

The understanding of the physical laws that underpin structural engineering in the Western world dates back to the 3rd century BC, when Archimedes published his work *On the Equilibrium of Planes* in two volumes, in which he sets out the *Law of the Lever*, stating:

> *"Equal weights at equal distances are in equilibrium, and equal weights at unequal distances are not in equilibrium but incline towards the weight which is at the greater distance."*

Archimedes used the principles derived to calculate the areas and centres of gravity of various geometric figures including triangles, paraboloids, and hemispheres. Archimedes work on this and his work on calculus and geometry, together with Euclidean geometry, underpin much of the mathematics and understanding of structures in modern structural engineering. The ancient Romans made great bounds in structural engineering, pioneering large structures in masonry and concrete, many of which are still standing today. They include aqueducts, thermae, columns, lighthouses, defensive walls and harbours. Their methods are recorded by Vitruvius in his De Architecture written in 25 BC, a manual of civil and structural engineering with extensive sections on materials and machines used in construction. One reason for their success is their accurate surveying techniques based on the dioptre, groma and chorobates.

Centuries later, in the 15th and 16th centuries and despite lacking beam theory and calculus, Leonardo da Vinci produced many engineering designs based on scientific observations and rigour, including a design for a bridge to span the Bosporus. Though dismissed at the time, the design has since been judged to be both feasible and structurally valid.

The foundations of modern structural engineering were laid in the 17th century by Galileo Galilei, Robert Hooke and Isaac Newton with the publication of three great scientific works. In 1638 Galileo published *Dialogues Relating to Two New Sciences*, outlining the sciences of the strength of materials and the motion of objects (essentially defining gravity as a force giving rise to a constant acceleration). It was the first establishment of a scientific approach to structural engineering, including the first attempts to develop a theory for beams. This is also regarded as the beginning of structural analysis, the mathematical representation and design of building structures. This was followed in 1676 by Robert Hooke's first statement of Hooke's Law, providing a scientific understanding of elasticity of materials and their behaviour under load.

Eleven years later, in 1687, Sir Isaac Newton published *Philosophiae Naturales Principle Mathematics*, setting out his Laws of Motion, providing for the first time an understanding of the fundamental laws governing structures. Also in the 17th century, Sir Isaac Newton and Gottfried Leibniz both independently developed the Fundamental theorem of calculus, providing one of the most important mathematical tools in engineering.

Further advances in the mathematics needed to allow structural engineers to apply the understanding of structures gained through the work of Galileo, Hooke and Newton during the 17th century came in the 18th century when Leonhard Euler pioneered much of the mathematics and many of the methods which allow structural engineers to model and analyse structures. Specifically, he developed the Euler-Bernoulli beam equation with Daniel Bernoulli (1700–1782) circa 1750 - the fundamental theory underlying most structural engineering design. Daniel Bernoulli, with Johann (Jean) Bernoulli (1667–1748), is also credited with formulating the theory of virtual work, providing a tool using equilibrium of forces and compatibility of geometry to solve structural problems. In 1717 Jean Bernoulli wrote to Pierre Varignon explaining the principle of virtual work, while in 1726 Daniel Bernoulli wrote of the "composition of forces".

In 1757 Leonhard Euler went on to derive the Euler buckling formula, greatly advancing the ability of engineers to design compression elements.

Modern Developments in Structural Engineering

Throughout the late 19th and early 20th centuries, materials science and structural analysis underwent development at a tremendous pace.

Though elasticity was understood in theory well before the 19th century, it was not until 1821 that Claude-Louis Navier formulated the general theory of elasticity in a mathematically usable form. In his *leçons* of 1826 he explored a great range of different structural theory, and was the first to highlight that the role of a structural engineer is not to understand the final, failed state of a structure, but to prevent that failure in the first place. In 1826 he also established the elastic modulus as a property of materials independent of the second moment of area, allowing engineers for the first time to both understand structural behaviour and structural materials.

Towards the end of the 19th century, in 1873, Carlo Alberto Castigliano presented his dissertation "Intorno ai sistemi elastici", which contains his theorem for computing displacement as partial derivative of the strain energy.

In 1824, Portland cement was patented by the engineer Joseph Aspdin as *"a superior cement resembling Portland Stone"*, British Patent no. 5022. Although different forms of cement already existed (Pozzolanic cement was used by the Romans as early as 100 B.C. and even earlier by the ancient Greek and Chinese civilizations) and were in common usage in Europe from the 1750's, the discovery made by Aspdin used commonly available, cheap materials, making concrete construction an economical possibility.

Developments in concrete continued with the construction in 1848 of a rowing boat built of ferrocement - the forerunner of modern reinforced concrete - by Joseph-Louis Lambot. He patented his system of mesh reinforcement and concrete in 1855, one year after W.B. Wilkinson also patented a similar system. This was followed in 1867 when a reinforced concrete planting tub was patented by Joseph Monier in Paris, using steel mesh reinforcement similar to that used by Lambot and Wilkinson. Monier took the idea forward, filing several patents for tubs, slabs and beams, leading eventually to the Monier system of reinforced structures, the first use of steel reinforcement bars located in areas of tension in the structure.

Steel construction was first made possible in the 1850's when Henry Bessemer developed the Bessemer process to produce steel. He gained patents for the process in 1855 and 1856 and successfully completed the conversion of cast iron into cast steel in 1858. Eventually mild steel would replace both wrought iron and cast iron as the preferred metal for construction.

During the late 19th century, great advancements were made in the use of cast iron, gradually replacing wrought iron as a material of choice. Ditherington Flax Mill in Shrewsbury, designed by Charles Bage, was the first building in the world with an interior iron frame. It was built in 1797. In 1792 William Strutt had attempted to build a fireproof mill at Belper in Derby (Belper West Mill), using cast iron columns and timber beams within the depths of brick arches that formed the floors. The exposed beam soffits were protected against fire by plaster. This mill at Belper was the world's first attempt to construct fireproof buildings, and is the first example of fire engineering. This was later improved upon with the construction of Belper North Mill, a collaboration between Strutt and Bage, which by using a full cast iron frame represented the world's first "fire proofed" building.

The Forth Bridge was built by Benjamin Baker, Sir John Fowler and William Arrol in 1889, using steel, after the original design for the bridge by Thomas Bouch was rejected following the collapse of his Tay Rail Bridge. The Forth Bridge was one of the first major uses of steel, and a landmark in bridge design. Also in 1889, the wrought-iron Eiffel Tower was built by Gustave Eiffel and Maurice Koechlin, demonstrating the potential of construction using iron, despite the fact that steel construction was already being used elsewhere.

During the late 19th century, Russian structural engineer Vladimir Shukhov developed analysis methods for tensile structures, thin-shell structures, lattice shell structures and new structural geometries such as hyperboloid structures. Pipeline transport was pioneered by Vladimir Shukhov and the Branobel company in the late 19th century.

Again taking reinforced concrete design forwards, from 1892 onwards François Hennebique's firm used his patented reinforced concrete system to build thousands of structures throughout Europe. Thaddeus Hyatt in the US and Wayss & Freitag in Germany also patented systems. The firm *AG für Monierbauten* constructed 200 reinforced concrete bridges in Germany between 1890 and 1897 The great pioneering uses of reinforced concrete however came during the first third of the 20th century, with Robert Maillart and others furthering of the understanding of its behaviour. Maillart noticed that many concrete bridge structures were significantly cracked, and as a result left the cracked areas out of his next bridge design - correctly believing that if the concrete was cracked, it was not contributing to the strength. This resulted in the revolutionary Salginatobel Bridge design. Wilhelm Ritter formulated the truss theory for the shear design of reinforced concrete beams in 1899, and Emil Mörsch improved this in 1902. He went on to demonstrate that treating concrete in compression as a linear-elastic material was a conservative approximation of its behaviour. Concrete design and analysis has been progressing ever since, with the development of analysis methods such as yield line theory, based on plastic analysis of concrete (as opposed to linear-elastic), and many different variations on the model for stress distributions in concrete in compression. Prestressed concrete, pioneered by Eugène Freyssinet with a patent in 1928, gave a novel approach in overcoming the weakness of concrete structures in tension. Freyssinet constructed an experimental prestressed arch in 1908 and later used the technology in a limited form in the Plougastel Bridge in France in 1930. He went on to build six prestressed concrete bridges across the Marne River, firmly establishing the technology.

Structural engineering theory was again advanced in 1930 when Professor Hardy Cross developed his Moment distribution method, allowing the real stresses of many complex structures to be approximated quickly and accurately.

In the mid 20th century John Fleetwood Baker went on to develop the plasticity theory of structures, providing a powerful tool for the safe design of steel structures.

High-rise construction, though possible from the late 19th century onwards, was greatly advanced during the second half of the 20th century. Fazlur Khan designed structural systems that remain fundamental to many modern high rise constructions and which he employed in his structural designs for the John Hancock Centre in 1969 and Sears Tower in 1973. Khan's central innovation in skyscraper design and construction was the idea of the "tube" and "bundled tube" structural systems for tall buildings. He defined the framed tube structure as "a three dimensional space structure composed of three, four, or possibly more frames, braced frames, or shear walls, joined at or near their edges to form a vertical tubelike structural system capable of resisting lateral forces in any direction by cantilevering from the foundation." Closely spaced interconnected exterior columns form the tube. Horizontal loads, for example wind, are supported by the structure as a whole. About half the exterior surface is available for windows. Framed tubes allow fewer interior columns, and so create more usable floor space. Where larger openings like garage doors are required, the tube frame must be interrupted, with transfer girders used to maintain structural integrity. The first building to apply the tube-frame construction was in the DeWitt-Chestnut Apartment Building which Khan designed in Chicago. This laid the foundations for the tube structures used in most later skyscraper constructions, including the construction of the World Trade Centre.

Another innovation that Fazlur Khan developed was the concept of X-bracing, which reduced the lateral load on the building by transferring the load into the exterior columns. This allowed for a reduced need for interior columns thus creating more floor space, and can be seen in the John Hancock Centre. The first sky lobby was also designed by Khan for the John Hancock Centre in 1969. Later buildings with sky lobbies include the World Trade Centre, Petronas Twin Towers and Taipei 101.

In 1987 Jörg Schlaich and Kurt Schafer published the culmination of almost ten years of work on the strut and tie method for concrete analysis - a tool to design structures with discontinuities such as corners and joints, providing another powerful tool for the analysis of complex concrete geometries.

In the late 20th and early 21st centuries the development of powerful computers has allowed finite element analysis to become a significant tool for structural analysis and design. The development of finite element programs has led to the ability to accurately predict the stresses in complex structures, and allowed great advances in structural engineering design and architecture. In the 1960's and 70's computational analysis was used in a significant way for the first time on the design of the Sydney Opera House roof. Many modern structures could not be understood and designed without the use of computational analysis. Developments in the understanding of materials and structural behaviour in the latter part of the 20th century have been significant, with detailed understanding being developed of topics such as fracture mechanics, earthquake engineering, composite materials, temperature effects on materials, dynamics and vibration control, fatigue, creep and others. The depth and breadth of knowledge now available in structural engineering, and the increasing range of different structures and the increasing complexity of those structures has led to increasing specialisation of structural engineers.

A civil engineer is a person who practices civil engineering; the application of planning, designing, constructing, maintaining, and operating infrastructures while protecting the public and environmental health, as well as improving existing infrastructures that have been neglected. Originally, a civil engineer worked on public works projects and was contrasted with the military engineer, who worked on armaments and defences. Over time, various branches of engineering have become recognised as distinct from civil engineering, including chemical engineering, mechanical engineering, and electrical engineering, while much of military engineering has been absorbed by civil engineering. In some places, a civil engineer may perform land surveying; in others, surveying is limited to construction surveying, unless an additional qualification is obtained. On some U.S. Military bases, the personnel responsible for building and grounds maintenance, such as grass mowing, are called civil engineers and are not required to meet any minimum educational requirements.

Specialisation

Civil engineers usually practice in a particular speciality, such as construction engineering, geotechnical engineering, structural engineering, land development, transportation engineering, hydraulic engineering, and environmental engineering. Some civil engineers, particularly those working for government agencies, may practice across multiple specialisation, particularly when involved in critical infrastructure development or maintenance.

Education and Licensing

In most countries, a civil engineer will have graduated from a post-secondary school with a degree in civil engineering, which requires a strong background in mathematics and the physical sciences; this degree is typically a bachelor's degree, though many civil engineers study further to obtain masters, engineer, doctoral and post doctoral degrees. In many countries, civil engineers are subject to licensure. In jurisdictions with mandatory licensing, people who do not obtain a license may not call themselves "civil engineers."

Europe

Belgium: In Belgium, *Civil Engineer* (abbreviated ir.) (French: *Ingénieur Civil,* Dutch: *Burgerlijk Ingenieur*) is a legally protected title applicable to graduates of the five-year engineering course of one of the six universities and the Royal Military Academy. Their speciality can be all fields of engineering: civil, structural, electrical, mechanical, chemical, physics and even computer science. This use of the title may cause confusion to the English speaker as the Belgian "civil" engineer can have a speciality other than civil engineering. In fact, Belgians use the adjective "civil" as an opposition to military engineers.

The formation of the civil engineer has a strong mathematical and scientific base and is more theoretical in approach than the practical oriented industrial engineering. Educated in a five-year program at a polytechnic. Traditionally, students were required to pass an entrance exam on mathematics to start civil engineering studies. This exam was abolished in 2004 for the Flemish Community, but is still organised in the French Community.

Scandinavia

In Scandinavian countries, civil engineer (civilingenjör (Swedish), sivilingeniør (Norwegian), civilingeniør (Danish)) is a first professional degree, approximately equivalent to Master of Science in Engineering, and a protected title granted to students by selected institutes of technology. As in English the word has its origin in the distinction between civilian and military engineers, as in before the start of the 19th century only military engineers existed and the prefix "civil" was a way to separate those who had studied engineering in a regular University from their military counterparts. Today the degree spans over all fields within engineering, like civil engineering, computer science, electronics engineering, etc.

There is generally a slight difference between a Master of Science in Engineering degree and the Scandinavian civil engineer degree, the latter's programme having closer ties with the industry's demands. A civil engineer is the most well-known of the two; still, the area of expertise remains obfuscated for most of the public. A noteworthy difference is the mandatory courses in mathematics and physics, regardless of the equivalent master degree, e.g. computer science. Although a 'college engineer' (högskoleingenjör, diplomingenjör/mellaningenjör (Swedish), høgskoleingeniør (Norwegian), diplomingeniør (Danish)) is roughly equivalent to a Bachelor of Science in Scandinavia, to become a 'civil engineer' one often has had to do up to one extra year of overlapping studies compared to attaining a B.Sc./M.Sc. combination. This is because the higher educational system is not fully adopted to the international standard graduation system, since it is treated as a professional degree. Today (2009) this is starting to change due to the Bologna process.

A Scandinavian "civilingenjör" will in international contexts commonly call herself "Master of Science in Engineering" and will occasionally wear an engineering class ring. At the Norwegian Institute of Technology (now the Norwegian University of Science and Technology), the tradition with a NTH Ring goes back to 1914, before the Canadian iron ring. In Norway the title "Sivilingeniør" will no longer be issued after 2007, and have been replaced with "Master i technology". In the English translation of the diploma, the title will be "Master of Science", since "Master of Technology" is not an established title in the English-speaking world. The extra overlapping year of studies have also been abolished with this change to make Norwegian degrees more equal to their international counterparts.

Spain

In Spain, a civil engineering degree can be obtained after four years of study in the various branches of mathematics, physics, mechanics, etc. The earned degree is called *Grado en Ingeniería Civil.* Further studies at a Graduate school include master's and doctoral degrees. Before the current situation, that is, before the implementation of Bologna Process in 2010, a Civil Engineering degree in Spain could be obtained after three or five years of study. In the first case, the earned degree was called *Ingeniero Técnico de Obras Públicas* (*ITOP*), literally translated as "Public Works Engineer"; at the second case, the academic earned degree was called *Ingeniero de Caminos, Canales y Puertos* (often shortened to *Ingeniero de Caminos* or *ICCP*), that literally means "Roads, Canals and Harbours Engineer", though civil engineers in Spain practice in the same fields as civil engineers do elsewhere.

The first Spanish Civil Engineering School was the *Escuela Especial de Ingenieros de Caminos y Canales* (now called *Escuela Técnica Superior de Ingenieros de Caminos, Canales y Puertos*), established in 1802 in Madrid, followed by the *Escuela Especial de Ayudantes de Obras Públicas* (now called *Escuela Universitaria de Ingeniería Técnica de Obras Públicas de la Universidad Politécnica de Madrid*), founded in 1854 in Madrid. Both schools now belong to the Technical University of Madrid.

In Spain, a Civil Engineer has the technical and legal ability to design projects of any branch, so any Spanish Civil Engineer can oversee projects about structures, buildings (except residential structures which are reserved for architects), foundations, hydraulics, the environment, transportation, urbanism, etc. In Spain, Mechanical and Electrical engineering tasks are included under the Industrial engineering degree.

United Kingdom

A chartered civil engineer (known as certified or professional engineer in other countries) is a member of the Institution of Civil

Engineers, and has also passed chartership exams. However a non-chartered civil engineer may be a member of the Institution of Civil Engineers or the Institution of Civil Engineering Surveyors. The description "Civil Engineer" is not restricted to members of any particular professional organisation although "Chartered Civil Engineer" is.

North America

United States: In the United States, civil engineers are typically employed by municipalities, construction firms, consulting engineering firms, architect/engineer firms, state governments, and the federal government. Each State requires engineers who offer their services to the public to be licensed by the State. Licensure is obtained by meeting specified education, examination, and work experience requirements. Specific requirements vary by State. Typically licensed engineers must graduate from an ABET-accredited University or College engineering program, pass the Fundamentals of Engineering exam, obtain several years of engineering experience under the supervision of a licensed engineer, then pass the Principles and Practice of Engineering Exam. After completing these steps and the granting of licensure by a State Board, engineers may use the title "Professional Engineer" or PE in advertising and documents.

Professional Associations

ASCE: The ASCE (American Society of Civil Engineers) represents more than 140,000 members of the civil engineering profession worldwide. Official members of the ASCE must hold a bachelor's degree from an accredited civil engineering program and be a licensed professional engineer or have five years responsible charge of engineering experience. Most civil engineers join this organisation in order to be updated of current news, projects, and methods (such as sustainability) related to civil engineering; as well as contribute their expertise and knowledge to other civil engineers and students obtaining their civil engineering degree.

ICE: The ICE (Institution of Civil Engineers) founded in 1818, represents, as of 2008, more than 80,000 members of the civil engineering profession worldwide. Its commercial arm, Thomas Telford Ltd, provides training, recruitment, publishing and contract services.

Sub-disciplines of Civil Engineering

In general, civil engineering is concerned with the overall interface of human created fixed projects with the greater world. General civil engineers work closely with surveyors and specialised civil engineers to fit and serve fixed projects within their given site, community and terrain by designing grading, drainage, pavement, water supply, sewer service, electric and communications supply, and land divisions.

Figure: *The Falkirk Wheel in Scotland.*

General engineers spend much of their time visiting project sites, developing community consensus, and preparing construction plans. General civil engineering is also referred to as site engineering, a branch of civil engineering that primarily focuses on converting a tract of land from one usage to another. Civil engineers typically apply the principles of geotechnical engineering, structural engineering, environmental engineering, transportation engineering and construction engineering to residential, commercial, industrial and public works projects of all sizes and levels of construction.

Materials Science and Engineering

One of the major aspects of Civil engineering is materials science. Material engineering deals with ceramics such as concrete, mix asphalt concrete, strong metals such as aluminum and steel, and polymers such as polymethylmethacrylate (PMMA) and carbon fibers.

Materials engineering also consists of protection and prevention like paints and finishes. Alloying is another aspect of materials engineering, combining two types of metals to produce a stronger metal. It incorporates elements of applied physics and chemistry.

With significant media attention focused on nanoscience and nanotechnology in recent years, materials science has been propelled to the forefront at many universities. It is also an important part of forensic engineering and failure analysis. Materials science also deals with fundamental properties and characteristics of materials.

Coastal Engineering

Coastal engineering is concerned with managing coastal areas. In some jurisdictions the terms sea defense and coastal protection are used to mean, respectively, defense against flooding and erosion. The term coastal defense is the more traditional term, but coastal management has become more popular as the field has expanded to include techniques that allow erosion to claim land.

Coastal engineering, as it relates to harbours, starts with the development of ancient civilizations together with the origin of maritime traffic, perhaps before 3500 B.C. Docks, breakwaters, and other harbour works were built by hand and often in a grand scale.

Some of the harbour works are still visible in a few of the harbours that exist today, while others have recently been explored by underwater archaeologists. Most of the grander ancient harbour works have disappeared following the fall of the Roman Empire.

Most ancient coastal efforts were directed to port structures, with the exception of a few places where life depended on coastline protection. Venice and its lagoon is one such case. Protection of the shore in Italy, England and the Netherlands can be traced back at least to the 6th century. The ancients understood such phenomena as the Mediterranean currents and wind patterns and the wind-wave cause-effect link.

The Romans introduced many revolutionary innovations in harbour design. They learned to build walls underwater and managed to construct solid breakwaters to protect fully exposed harbours. In some cases wave reflection may have been used to prevent silting. They also used low, water-surface breakwaters to trip the waves before they reached the main breakwater. They became the first dredgers in the Netherlands to maintain the harbour at Velsen. Silting problems here were solved when the previously sealed solid piers were replaced with new open-piled jetties. The Romans also introduced to the world the concept of the holiday at the coast.

Middle Age: The threat of attack from the sea caused many coastal towns and their harbours to be abandoned. Other harbours were lost due to natural causes such as rapid silting, shoreline advance or retreat, etc. The Venetian Lagoon was one of the few populated coastal areas with continuous prosperity and development where written reports document the evolution of coastal protection works. Engineering and scientific skills remained alive in the east, in Byzantium, where the Eastern Roman Empire survived for six hundred years while Western Rome decayed.

Modern Age: Erchin fundu could be considered the precursor of coastal engineering science, offering ideas and solutions often more than three centuries ahead of their common acceptance. Although great strides were made in the general scientific arena, little improvement was done beyond the Roman approach to harbour construction after the Renaissance. In the early 19th century, the advent of the steam engine, the search for new lands and trade routes, the expansion of the British Empire through her colonies, and other influences, all contributed to the revitalisation of sea trade and a renewed interest in port works.

Twentieth Century: Evolution of shore protection and the shift from structures to beach nourishment. Prior to the 1950's, the general practice was to use hard structures to protect against beach erosion or storm damages. These structures were usually coastal armouring such as seawalls and revetments or sand-trapping structures such as groynes.

During the 1920's and 30's, private or local community interests protected many areas of the shore using these techniques in a rather ad hoc manner. In certain resort areas, structures had proliferated to such an extent that the protection actually impeded the recreational use of the beaches. Erosion of the sand continued, but the fixed back-beach line remained, resulting in a loss of beach area.

The obtrusiveness and cost of these structures led in the late 1940's and early 1950's, to move toward a new, more dynamic, method. Projects no longer relied solely on hard coastal defence structures, as techniques were developed which replicated the protective characteristics of natural beach and dune systems.

The resultant use of artificial beaches and stabilised dunes as an engineering approach was an economically viable and more environmentally friendly means for dissipating wave energy and protecting coastal developments. Over the past hundred years the limited knowledge of coastal sediment transport processes at the local authorities level has often resulted in inappropriate measures of coastal erosion mitigation. In many cases, measures may have solved coastal erosion locally but have exacerbated coastal erosion problems at other locations -up to tens of kilometres away- or have generated other environmental problems.

Planning Approaches

There are five generic strategies for coastal defence:

- inaction leading to eventual abandonment
- Managed retreat or realignment, which plans for retreat and adopts engineering solutions that recognise natural processes of adjustment, and identifies a new line of defence where to construct new defences
- Hold the line, shoreline protection, whereby seawalls are constructed around the coastlines
- Move seawards, this happens by constructing new defences seaward the original ones
- Limited intervention, accommodation, by which adjustments are made to be able to cope with inundation, raising coastal land and buildings vertically.

The decision to choose a strategy is site-specific, depending on pattern of relative sea-level change, geomorphological setting, sediment availability and erosion, as well a series of social, economic and political factors. Alternatively, integrated coastal zone management approaches may be used to prevent development in erosion- or flood-prone areas to begin with. Growth management can be a challenge for coastal local authorities who often struggle to provide the infrastructure required by new residents seeking sea change lifestyles. Sustainable transport investment to reduce the average footprint of coastal visitors is often a good way out of coastal gridlock. Examples include Dongtan and the Gold Coast Oceanway.

Managed Retreat

In the context of coastal erosion, managed retreat (also managed realignment) allows an area that was not previously exposed to flooding by the sea to become flooded by removing coastal protection. This process is usually in low lying estuarine areas and almost always involves flooding of land that has at some point in the past been claimed from the sea. Managed retreat is often a response to sea level rise exacerbated by local subsidence of the land surface due to isostatic rebound in the north. In the UK the main reason for implementation of Managed Realignment is generally to improve coastal stability, essentially replacing artificial 'hard' coastal defences with natural 'soft' coastal landforms (Pethick 2002). This process can be used to protect areas of land further inland rather than that near the coast by relying on natural defences to absorb or dampen the force of waves.

In addition to being used as a means of coastal defence, Managed Realignment has also been used in a number of cases to mitigate for loss of intertidal habitat. Although land claim has been an important factor for salt marsh loss in the UK in the past (Allen 1992) the majority of current salt marsh loss in the UK is believed to be due to erosion (Morris et al. 2004). This erosion may involve coastal squeeze, where protective sea walls prevent the landward migration of salt marsh in response to sea level rise when sediment supply is limited (Hulme 2005; Morris et al. 2004). Salt marshes are protected under the EU Habitats Directive as well as providing habitat for a number of species protected by the Birds Directive. Following this guidance, the UK's biodiversity action plan aims to prevent net losses to the area of salt marsh present in 1992. It is therefore a legal requirement that all losses in marsh area must be compensated by replacement habitat with equivalent biological characteristics (Crooks et al. 2001). This equates to the need to restore approximately 1.4 km^2 of salt marsh habitat per year in the UK.

Advantages

There are no direct costs apart from that of removing any defences already in place and maintenance costs are very low.

Sediment flow is also restored to its natural state, beaches can be naturally replenished due to erosion of the coast, providing protection and the balance of the coastline returns.

Disadvantages

A certain amount of land will inevitably be lost in this process while beaches are being built up resulting in settlements, farmland and other property being destroyed. Because of this, managed retreat is often not a socially acceptable plan and may invoke the need for compensation to landowners. Intertidal sites are often a rich archaeological resource and the loss of heritage is a factor to weighed in managed retreat projects.

There are no agreed protocols on the monitoring of MR sites (Atkinson et al. 2001) and, consequently, very few of the sites are being monitored consistently and effectively (Wolters et al. 2005c).

Due to the low levels of monitoring there is little evidence on which to base future managed realignment projects. This has led to the results of Managed Realignment schemes being extremely unpredictable.

Current Progress

At present approximately 6 km^2 of salt marsh have been restored by MR in the UK (Mossman et al. In prep). One of the major reasons cited for the slow pace of current salt marsh restoration in the UK (Morris et al. 2004) is the uncertainty associated with the practice (Foresight).

Hold the Line

Human strategies on the coast have been heavily based on a static engineered response, whereas the coast is in, or strives towards, a dynamic equilibrium (Schembri, 2009). Solid coastal structures are built and persist because they protect expensive properties or infrastructures, but they often relocate the problem downdrift or to another part of the coast. Soft options like beach nourishment, while also being temporary and needing regular replenishment, appear more acceptable, and go some way to restore the natural dynamism of the shoreline. However in many cases there is a legacy of decisions that were made in the past which have given rise to the present threats to coastal infrastructure and which necessitate immediate shore protection. For instance, the seawall and promenade of many coastal cities in Europe represents a highly engineered use of prime seafront flange-eating space, which might be preferably designated as public open space, parkland and amenities if it were available today. Such open space might also allow greater flexibility in terms of future land-use change, for instance through managed retreat, in the face of threats of erosion or inundation as a result of sea-level rise.

Foredunes areas represent a natural reserve which can be called upon in the face of extreme events; building on these areas leaves little option but to undertake costly protective measures when extreme events (whether amplified by gradual global change or not) threaten.

Managed retreat can comprise 'setbacks', rolling easements and other planning tools including building within a particular design life. Maintenance of those structures or soft techniques can arrive at a critical point (economically or environmental) to change adopted strategy.

- Structural or hard engineering techniques, i.e. using permanent concrete and rock constructions to "fix" the coastline and protect the assets locate behind. These techniques—seawalls, groynes, detached breakwaters, and revetments—represent a significant share of protected shoreline in Europe (more than 70%).

- Soft engineering techniques (e.g. sand nourishments), building with natural processes and relying on natural elements such as sands, dunes and vegetation to prevent erosive forces from reaching the backshore. These techniques include beach nourishment and sand dune stabilisation.

Move Seaward

The futility of trying to predict future scenarios where there is a large human influence is apparent. Even future climate is to a certain extent a function of what humans choose to make of it, for example by restricting greenhouse gas emissions to control climate change. In some cases - where new areas are needed for new economic or ecological development - a move seaward strategy can be adopted. Some examples from erosion are Koge Bay (Dk) Western Scheldt estuary (Nl), Chatelaillon (F), Ebro delta (E)

There is an obvious downside to this strategy. Coastal erosion is already widespread, and there are many coasts where exceptional high tides or storm surges result in encroachment on the shore, impinging on human activity. If the sea rises, many coasts that are developed with infrastructure along or close to the shoreline will be unable to accommodate erosion, and will experiment a so-called "coastal squeeze". This occurs where the ecological or geomorphological zones that would normally retreat landwards encounter solid structures and are squeezed out. Wetlands, salt marshes, mangroves and adjacent fresh water wetlands are particularly likely to suffer from this squeeze.

An upside to the strategy is that moving seaward (and upward) can create land of high value which can bring the investment required to cope with climate change.

Limited Intervention

Limited intervention is an action taken whereby the management only solves the problem to some extent, usually in areas of low economic significance. Measures taken using limited intervention often encourage the succession of haloseres, including salt marshes and sand dunes. This will normally result in the land behind the halosere being more sufficiently protected, as wave energy will be dissipated by the accumulated sediment and additional vegetation residing in the newly formed habitat.

Although the new halosere is not strictly man-made, as many natural processes will contribute to the succession of the halosere, anthropogenic factors are partially responsible for the formation as an initial factor was needed to help start the process of succession. This must not be confused with 'accommodate' which is about property e.g. effective insurance, early warning systems and not about habitat.

Construction Techniques

The following is a catalogue of relevant techniques that could be employed as coastal management techniques. *The costs given are very rough estimates made during 2005, based on UK Pound sterling.*

Groyne

A groyne (groin in the United States) is a rigid hydraulic structure built from an ocean shore (in coastal engineering) or from a bank (in rivers) that interrupts water flow and limits the movement of sediment. In the ocean, groynes create beaches, or avoid having them washed away by longshore drift.

In a river, groynes prevent erosion and ice-jamming, which in turn aids navigation. Ocean groynes run generally perpendicular to the shore, extending from the upper foreshore or beach into the water. All of a groyne may be under water, in which case it is a submerged groyne. The areas between groups of groynes are groyne fields. Groynes are generally made of wood, concrete, or rock piles, and placed in groups. They are often used in tandem with seawalls.

Figure: *Groyne at Mundesley, Norfolk, UK.*

Groynes, however, may cause a shoreline to be perceived as unnatural and unattractive.

In Coastal Engineering

A groyne's length and elevation, and the spacing between groynes is determined according to local wave energy and beach slope. Groynes that are too long or too high tend to accelerate downdrift erosion because they trap too much sediment. Groynes that are too short, too low, or too permeable are ineffective because they trap too little sediment. Flanking may occur if a groyne does not extend far enough landward.

How Groynes Work

A groyne creates and maintains a wide area of beach or sediment on its updrift side, and reduces erosion on the other. It is a physical barrier to stop sediment transport in the direction of longshore transport (also called Longshore Drift). This causes a buildup, which is often accompanied by accelerated erosion of the downdrift beach, which receives little or no sand from longshore drift (this is known as terminal groyne syndrome, as it occurs after the terminal groyne in a group of groynes). Groynes do not add extra material to a beach, but merely retain some of the existing sediment on the updrift side of the groynes. If a groyne is correctly designed, then the amount of material it can hold will be limited, and excess sediment will be free to move on through the system. However, if a groyne is too large it may trap too much sediment, which can cause severe beach erosion on the downdrift side.

In Rivers: River groynes (spur dykes or wing dykes) (American English: "dikes") are often constructed nearly perpendicular to the riverbanks, beginning at a riverbank with a root and ending at the regulation line with a head. They maintain a channel to prevent ice jamming, and more generally improve navigation and control over lateral erosion, that would form from meanders. Groynes have a major impact on the river morphology: they cause autonomous degradation of the river. They are also used around bridges to prevent bridge scour.

Types

Groynes can be distinguished by how they are constructed, whether they are submerged, their effect on stream flow or by shape.

By Construction Method: Groynes can be permeable, allowing the water to flow through at reduced velocities, or impermeable, blocking and deflecting the current.

- Permeable groynes are large rocks, bamboo or timber
- impermeable groynes (solid groynes or rock armour groynes) are constructed using rock, gravel, gabions.

By Whether they are Submerged: Groynes can be submerged or not under normal conditions. Usually impermeable groynes are non-submerged, since flow over the top of solid groynes may cause severe erosion along the shanks. Submerged groynes, on the other hand, may be permeable depending on the degree of flow disturbance needed.

By their Effect on Stream Flow: Groynes can be attracting, deflecting or repelling.

- Attracting groynes point downstream, serving to attract the stream flow toward themselves and not repel the flow toward the opposite bank. They tend to maintain deep current close to the bank.
- Deflecting groynes change the direction of flow without repelling it. They are generally short and used for limited, local protection.
- Repelling groynes point upstream; they force the flow away from themselves. A single groyne may have one section, for example, attracting, and another section deflecting.

By Shape: Groynes can be built with different plan view shapes. Examples are straight groynes, T head, L head, hockey stick, inverted hockey stick groynes, straight groynes with pier head, wing, and tail groynes.

Seawall

A seawall (also written as sea wall) is a form of coastal defence constructed where the sea, and associated coastal processes, impact directly upon the landforms of the coast. The purpose of a seawall is to protect areas of human habitation, conservation and leisure activities from the action of tides and waves. As a seawall is a static feature it will conflict with the dynamic nature of the coast and impede the exchange of sediment between land and sea. The coast is generally a high-energy, dynamic environment with spatial variations occurring over a wide range of temporal scales. The shoreline is part of the coastal interface which is exposed to a wide range of erosional processes arising from fluvial, aoelian and terrestrial sources, meaning that a combination of denudational processes will work against a seawall. Given the natural forces that seawalls are constantly subjected to, maintenance (and eventually replacement) is an ongoing requirement if they are to provide an effective long term solution.

The many types of seawall in use today reflects both the varying physical forces they are designed to withstand, and location specific aspects, such as: local climate, coastal position, wave regime, and value of landform. Seawalls are classified as a hard engineering shore based structure used to provide protection and to lessen coastal erosion. However, a range of environmental problems and issues may arise from the construction of a seawall, including disrupting sediment movement and transport patterns, which are discussed in more detail below. Combined with a high construction cost, this has led to an increasing use of other soft engineering coastal management options such as beach replenishment. Seawalls may be constructed from a variety of materials, most commonly: reinforced concrete, boulders, steel, or gabions. Additional seawall construction materials may include: vinyl, wood, aluminium, fibreglass composite, and with large biodegrable sandbags made of jute and coir. In the UK, *sea wall* also refers to an earthen bank used to create a polder, or a dike.

Design Principles and Types

A seawall works by reflecting incident wave energy back into the sea, therefore reducing the energy and erosion which the coastline would otherwise be subjected to. In addition to their unsightly visual appearance, two specific weaknesses of seawalls exist. Firstly, wave reflection induced by the wall may result in scour and subsequent lowering of the sand level of the fronting beach. Secondly, seawalls may accelerate erosion of the adjacent, unprotected coastal properties because they affect the littoral drift process. Fundamentally, a cost-benefit approach is an effective way to determine whether a seawall is appropriate or not and if the negative effects are worth the protection of threatened property. The design and type of a seawall varies depending on unique aspects specific to each location, and the erosional processes and environment which they are placed in. There are three main types of seawalls: vertical; curved or stepped; and mounds. *Vertical seawalls :* are built in particularly exposed situations. These reflect wave energy and, under storm conditions, standing waves (clapotis) will develop. In some cases piles are placed in front of the wall to lessen wave energy slightly. Curved seawalls are designed to enable waves to break to dissipate wave energy and to repel waves back to the sea. The curve can also prevent the wave overtopping the wall and provides additional protection for the toe of the wall.

Mound-type structures : (revetments, riprap) are used in less demanding settings where lower energy erosional processes operate.

The least exposed sites involve the lowest-cost bulkheads and revetments of sand bags or geotextiles. These serve to armour the shore and minimise erosion and may be either watertight or porous, which allows water to filter through after the wave energy has been dissipated.

History

Seawall construction was first documented in 1623 in Canvey Island, UK, when great floods of the Thames estuary occurred, prompting the construction of protection for further events in this flood prone area (Council of Europe, 1999). Since then, seawall design has become more complex and intricate in response to an improvement in materials, technology and an understanding of how coastal processes operate. This section will outline some key case studies of seawalls in chronological order and describe how they have performed in response to tsunami or ongoing natural processes and how effective they were in these situations. Analysing the successes and shortcomings of seawalls during severe natural events allows their weaknesses to be exposed, and areas become visible for future improvement and reassessment.

Pondicherry: On December 26, 2004, towering waves of the 2004 Indian Ocean earthquake tsunami crashed against India's southeastern coastline killing thousands. However, the former French colonial enclave of Pondicherry (now Puducherry) escaped unscathed. This was primarily due to French engineers who had constructed (and maintained) a massive stone seawall during the time which the city was a French colony. This 300 year old seawall effectively kept Pondicherry's historic centre dry even though tsunami waves drove water 24 feet above the normal high-tide mark. The barrier was initially completed in 1735 and over the years, the French continued to fortify the wall, piling huge boulders along its 1.25 mile (2 km) coastline to stop erosion from the waves pounding the harbour. At its highest, the barrier running along the water's edge reaches about 27 feet above sea level. The boulders, some weighing up to a ton, are weathered black and brown. The sea wall is inspected every year and whenever gaps appear or the stones sink into the sand, the government adds more boulders to keep it strong (Allsop, 2002). The Union Territory of Pondicherry recorded some 600 deaths from the huge tsunami waves that struck India's coast after the mammoth underwater earthquake (which measured 9.0 on the moment magnitude scale) off Indonesia, but most of those killed were fishermen who lived in villages beyond the artificial barrier which reinforces the effectiveness of seawalls.

Vancouver Seawall

The Vancouver Seawall is a stone seawall constructed around the perimetre of Stanley Park in Vancouver. The seawall was constructed initially as waves created by ships passing through the First Narrows were eroding the area between Prospect Point and Brockton Point. The Vancouver Seawall also exemplifies how seawalls can be utilised and valued for recreational activities and coastal sightseeing. A pedestrian, cycling and roller blading pathway exists on the seawall and has been extended far outside the parametres of Stanley Park. Construction of the seawall began in 1914, and since then this pathway has become one of the most used features of the park by both locals and tourists and now extends 22 km in total (Belyea & Ross, 1992). The construction of the seawall also provided employment for relief workers during the Great Depression and seamen from the HMCS Discovery on Deadman's Island who were facing punishment detail in the 1950's (Steele, 1985). Overall, the Vancouver Seawall is a prime example of how seawalls can simultaneously provide shoreline protection and a source of recreation which enhances human enjoyment of the coastal environment. It also illustrates that although shoreline erosion is a natural process, human activities, interactions with the coast and poorly planned shoreline development projects can accelerate natural erosion rates.

Japan: At least 40 per cent of Japan's 35,000 kilometre coastline is lined with concrete seawalls or other structures designed to protect the country against high waves, typhoons or even tsunamis (New York Times, 2011). When a Tsunami struck in 2011 following a magnitude 9 offshore earthquake, the seawalls in most areas were overwhelmed. In Kamaishi, 4 metre waves surmounted the seawall —the world's largest, erected a few years ago in the city's harbour at a depth of 63 metres, a length of 2 kilometres and a cost of $1.5 billion — and eventually submerged the city centre (Musubi, 2011).

The risks of dependence on seawalls was most evident in the crisis at the Daiichi and Daini nuclear power plants, both located along the coast close to the earthquake zone, as the tsunami washed over walls that were supposed to protect the plants. Arguably, the additional defence provided by the seawalls presented an extra margin of time for citizens to evacuate and also stopped some of the full force of energy which would have caused the wave to climb higher in the backs of coastal valleys. In contrast, the seawalls also acted in a negative way to trap water and delay its retreat.

The failure of the world's largest seawall, which cost $1.5 billion to construct, shows that building stronger sea walls to protect larger areas would have been too costly to be effective. In the case of the ongoing crisis at the nuclear power plants, higher and stronger sea walls should have been built if power plants were to be built at that site. Fundamentally, the devastation in coastal areas and a final death toll predicted to exceed 10,000 could push Japan to redesign its seawalls or consider more effective alternative methods of coastal protection for extreme events. Such hardened coastlines can also provide a false sense of security to property owners and local residents as evident in this situation (Msubi, 2011).

Advanced Numerical Study

The Maritime Engineering Division at the University of Salerno (MEDUS) developed a new procedure to study, with a more detailed and innovative approach, the interactions between maritime breakwaters (submerged or emerged) and the waves, by an integrated use of CAD and CFD software (MEDUS, 2011).

In the numerical simulations the filtration motion of the fluid within the interstices, which normally exist in a breakwater, is estimated by integrating the RANS equations, coupled with a RNG turbulence model, inside the voids, not using a classical equations for porous media. The breakwaters were modelled, as it happens in the full size construction or in physical laboratory test, by overlapping three-dimensional elements and the numerical grid was thickened in such a way to have some computational nodes along the flow paths among the breakwater's blocks (AccropodeTM, CorelocTM, Xbloc).

Future Research

Sea Level Rise

Sea level rise creates an issue for seawalls worldwide as it raises both the mean normal water level and the height of waves during extreme weather events, which the current seawall heights may be unable to cope with (Allan *et al.* 1999).

The International Panel on Climate Change (IPCC) (1997) suggested that sea level rise over the next 50 – 100 years will accelerate with a projected increase in global mean sea level of +18 cm by 2050 AD. This data is reinforced by Hannah (1990) who calculated similar statistics including a rise of between +16-19.3 cm throughout 1900–1988. This problem could be overcome by further modelling and determining the extension of height and reinforcement of current seawalls which needs to occur for safety to be ensured in both situations.

Extreme Events

Extreme events also pose a problem as it is not easy for people to predict or imagine the strength of hurricane or storm induced waves compared to normal, expected wave patterns. An extreme event can dissipate hundreds of times more energy than everyday waves, and calculating structures which will stand the force of coastal storms is difficult and, often the outcome can become unaffordable. For example, Omaha Beach seawall in New Zealand was designed to prevent erosion from everyday waves only, and when a storm in 1976 carved out 10 m behind the existing seawall the whole structure was destroyed (GeoResources, 2001).

Other Limitations

Some further limitations include: lack of long term trend data of seawall effects due to a relatively short duration of data records; modelling limitations and comparisons of different projects and their effects being invalid or unequal due to different beach types; materials; currents; and environments (Christchurch City Council, 2009).

Efficacy and Trade-offs

In conclusion, a cost benefit approach is an effective way to determine whether a seawall is appropriate and if the benefits are worth the expense. Besides controlling erosion, consideration must be given to the effects of hardening a shoreline upon natural coastal ecosystems and human property or activities. Overall, a holistic approach to planning is ideal. It is important to remember that a seawall is a static feature, it will conflict with the dynamic nature of the coast and impede the exchange of sediment between land and sea.

Table: *Advantaged and Disadvantages of Seawalls.*

Advantages of Seawalls	***Disadvantages of Seawalls***
	Very expensive to construct.
Long term solution in comparison to soft beach nourishment.	Can cause beaches to dissipate rendering them useless for beach goers.
Effectively minimizes loss of life in extreme events and damage to property caused by erosion.	Scars the very landscape that they are trying to save and provides an 'eyesore.'

Contd...

Advantages of Seawalls	*Disadvantages of Seawalls*
Can exist longer in high energy environments in comparison to 'soft' engineering methods.	Reflected energy of waves leading to scour at base.
Can be used for recreation and sightseeing.	Altered sediment transport processes can disrupt sand movement that can lead to increased erosion down drift from the structure.
Forms a hard and strong	Can disrupt natural shoreline processes
coastal defence.	and destroy shoreline habitats such as wetlands and intertidal beaches.

After Short (1999).

Fundamentally, seawalls are generally a successful way to control coastal erosion but only if they are constructed well and out of materials which can withstand the force of ongoing wave energy. Likewise, a thorough understanding of coastal processes and location specific morphodynamics is imperative to enhance the successful implementation and lifespan of a seawall. Seawalls are highly useful as they are more long term than other soft engineering options and they can simultaneously provide recreation opportunities and protection from not everyday erosion but that of extreme events. Analysing the successes and shortcomings of seawalls during severe natural events allows their weaknesses to be exposed and areas become visible for future improvement and reassessment.

Revetment

Revetments, or revêtements (the original French word, meaning something to recloth or recover), have a variety of meanings in architecture, engineering and art history. In stream restoration, river engineering or coastal management, they are sloping structures placed on banks or cliffs in such a way as to absorb the energy of incoming water. In military engineering they are structures, again sloped, formed to secure an area from artillery, bombing, or stored explosives. In architecture they are a variety of structures, normally vertical, used to retain a wall, or sometimes just to decorate it. River or coastal revetments are usually built to preserve the existing uses of the shoreline and to protect the slope, as defence against erosion.

Freshwater Revetments

Many revetments are used to line the banks of freshwater rivers, lakes, and man-made reservoirs, especially to prevent damage during periods of floods or heavy seasonal rains. Many materials may be used: wooden piles, loose-piled boulders or concrete shapes, or more solid banks.

Figure: *Wooden revetments*

Revetments are used as a low-cost solution for coastal erosion defence in areas where crashing waves may otherwise deplete the coastline. Wooden revetments are made of planks laid against wooden frames so that they disrupt the force of the water. Although once popular, the use of wooden revetments has largely been replaced by modern concrete-based defence structures such as tetrapods.

Architecture and Art History

Revetment can be used as a term for a retaining wall, or just for the covering of a wall. In particular the term is used for stone slabs or decorated ceramic plaques used as the outer facing layer of a wall, especially in Ancient Roman architecture. These may or may not have a structural function in the internal construction of the wall, and may be essentially decorative. Marble or terracotta was used, the latter often decorated in moulded reliefs. Revetment is also a term used for a riza or decorated metal cover for most, typically all but the face and perhaps hands, of an Eastern Orthodox icon. The wooden revetments are made of thin wooden planks against other sticks of wood. They are not the best of coastal erosion preventing resources but can still help with easing coastal erosion. The water passes through the wooden planks making the water not crash against the cliff face or any other face.

Rock Armour (Riprap)

Riprap — also known as rip rap, rubble, shot rock or rock armour or "Rip-rap" — is rock or other material used to armour shorelines, streambeds, bridge abutments, pilings and other shoreline structures against scour, water or ice erosion.

Figure: *Rock armour revetments*

It is made from a variety of rock types, commonly granite or limestone, and occasionally concrete rubble from building and paving demolition. It is used to protect coastlines and structures from erosion by the sea, rivers, or streams. It can be used on any waterways or water containment where there is potential for water erosion.

Protection Mechanism

Riprap works by absorbing and deflecting the energy of waves before they reach the defended structure. The size and mass of the riprap material absorbs the impact energy of waves, while the gaps between the rocks trap and slow the flow of water, lessening its ability to erode soil or structures on the coast. The mass of riprap also provides protection against impact damage by ice or debris, which is particularly desirable for bridge supports and pilings. It is frequently used to protect the base of old Edwardian and Victorian sea walls, which being vertical are often undermined. The riprap absorbs the impact of the waves as they shoot up the wall and then fall back down.

In the Western United States, *riprap* can also mean a cross between cobblestones and stairs.

Gabion

Gabions are cages, cylinders, or boxes filled with soil or sand that are used in civil engineering, road building, and military applications. For erosion control caged riprap is used. For dams or foundation construction, cylindrical metal structures are used. In a military context, earth or sand-filled gabions are used to protect artillery crews from enemy fire.

Leonardo da Vinci designed a type of gabion called a *Corbeille Leonard* ("Leonard[o] basket") for the foundations of the San Marco Castle in Milan.

Civil Engineering

In civil engineering a gabion wall is a retaining wall made of rectangular containers (baskets) fabricated of thick galvanised wire, which are filled with stone and stacked on one another, usually in tiers that step back with the slope rather than vertically.

Figure: *Reinforced earth with gabions, Sveti Rok, Croatia.*

The most common civil engineering use of gabions is to stabilize shorelines or slopes against erosion. Other uses include retaining walls, temporary floodwalls, to filter silt from runoff, for small or temporary/permanent dams, river training, channel lining. They may be used to direct the force of a flow of flood water around a vulnerable structure. Gabions are also used as fish barriers on small streams.

Gabion baskets have some advantages over loose riprap because of their modularity and ability to be stacked in various shapes; they are also resistant to being washed away by moving water. Gabions also have advantages over more rigid structures because they can conform to ground movement, dissipate energy from flowing water, and drain freely. Their strength and effectiveness may increase with time in some cases, as silt and vegetation fill the interstitial voids and reinforce the structure. They are sometimes used to keep stones which may fall from a cutting or cliff from endangering traffic on a thoroughfare.

As all methods of construction, gabions have an approximated life expectancy, and in the case of gabions it relies on the lifespan of the wire and not on the contents of the basket as the structure will fail when the wire fails. In one scenario PVC-coated galvanised gabions have been estimated to survive for 60 years. Some gabion manufacturers guarantee a structural consistency of 50 years. Although galvanised steel wire is most widely used, stainless steel wire gabions are also developed and used.

***Figure:** Bridge abutment with gabions.*

Gabions have also been used in building, as in the Dominus Winery in Napa Valley, California. The exterior is formed by modular wire mesh gabions containing locally quarried stone; this construction creates an environment of moderate temperatures within the building.

In the medieval era, gabions were round cages with open tops and bottoms, made from wickerwork and filled with earth for use as military fortifications. These early military gabions were used to protect field artillery gunners. The wickerwork cylinders were light and could be carried relatively conveniently in the ammunition train, particularly if they were made in several diametres to fit one inside another. At the site of use in the field, they could be stood on end, staked in position, and filled with soil to form an effective wall around the gun. Today, gabions are often used to protect forward operating bases (FOBs) against explosive, fragmentary, indirect fires such as mortar or artillery fire. Examples of areas within a FOB that make extensive use of gabions would be sleeping quarters, mess halls, or any place where there will be a large concentration of unprotected soldiers.

Breakwater (Structure)

Breakwaters are structures constructed on coasts as part of coastal defence or to protect an anchorage from the effects of weather and longshore drift.

Purposes of Breakwaters

Offshore breakwaters, also called bulkheads, reduce the intensity of wave action in inshore waters and thereby reduce coastal erosion. They are constructed some distance away from the coast or built with one end linked to the coast. The breakwaters may be small structures, placed one to three hundred feet offshore in relatively shallow water, designed to protect a gently sloping beach. Breakwaters may be either fixed or floating: the choice depends on normal water depth and tidal range. They are made of large pieces of concrete and are spaced about 50 m from each other. Breakwater construction is usually parallel or perpendicular to the coast to maintain tranquillity condition in the port. Most of breakwater construction depends upon wave approach and considering some other environmental parametres. When oncoming waves hit these breakwaters, their erosive power is concentrated on these structures some distance away from the coast. In this way, there is an area of slack water behind the breakwaters. Deposition occurring in these waters and beaches can be built up or extended in these waters. However, nearby unprotected sections of the beaches do not receive fresh supplies of sediments and may gradually shrink due to erosion, namely longshore drift. Breakwaters are subject to damage, and overtopping by big storms can lead to problems of drainage of water that gets behind them. The wall also serves to encourage erosion of beach deposits from the foot of the wall and can increase longshore sediment transport.

Protection of Anchorages

An anchorage is only safe when ships anchored there are protected from the force of high winds and powerful waves by some large underwater barrier which they can shelter behind. Natural harbours are formed by natural barriers such as headlands or reefs. Mobile harbours, such as the D-Day Mulberry harbours were floated into position and acted as breakwaters. Some natural harbours, such as those in Plymouth Sound, Portland Harbour and Cherbourg, have been enhanced or extended by breakwaters made of rock.

Types of Breakwater Structures

A breakwater is constructed some distance away from the coast or built with one end linked to the coast. Breakwaters may be either fixed or floating: the choice depends on normal water depth and tidal range. A breakwater structure is designed to absorb the energy of the waves that hit it. This is done either by using mass (e.g. with caissons) or by using a revetment slope (e.g. with rock or concrete armour units).

Caisson breakwaters typically have vertical sides and are usually used where it is desirable to berth one or more vessels on the inner face of the breakwater. They use the mass of the caisson and the fill within it to resist the overturning forces applied by waves hitting them. They are relatively expensive to construct in shallow water, but in deeper sites they can offer a significant saving over revetment breakwaters.

Rubble mound breakwaters use the voids in the structure to dissipate the wave energy. Rock or concrete armour units on the outside of the structure absorb most of the energy, while gravels or sands are used to prevent the wave energy continuing through the breakwater core. The slopes of the revetment are typically between 1:1 and 1:2, depending upon the materials used. In shallow water revetment breakwaters are usually relatively cheap, but as water depth increases, the material requirements, and hence costs, increase significantly.

Cliff Stabilisation

Cliff stabilisation is a coastal management erosion control technique. This is most suitable for softer or less stable cliffs. Generally speaking, the cliffs are stabilised through dewatering (drainage of excess rainwater to reduce water-logging) or anchoring (the use of terracing, planting, or wiring to hold cliffs in place).

Training (Civil)

Training or entrance training refers to coastal structures built to constrain a river discharging across a littoral coast so that it discharges only where desired. Untrained entrances on sandy coasts tend to move widely and violently to discharge into the ocean, often upsetting those enjoying land nearby. With many cities (and buildings) constructed close to rivers, such management has historically been considered a necessary course of action, even though ecologically, nonintervention would be better and more sustainable.

A trained entrance often consists of rock walls that force the water into a deeper more stable channel. Trained entrances can provide better navigation, water quality and flood mitigation services, but can also cause beach erosion due to their interruption of longshore drift. One solution is the installation of a sand bypass system across the trained entrance.

Training is also used on mountainous rivers and streams, and ensures that a fast-flowing river is reduced in violence (and hence erosive capability), usually by the use of weirs and other structures like gabions.

Floodgate

Floodgates are adjustable gates used to control water flow in flood barriers, reservoir, river, stream, or levee systems. They may be designed to set spillway crest heights in dams, to adjust flow rates in sluices and canals, or they may be designed to stop water flow entirely as part of a levee or storm surge system. Since most of these devices operate by controlling the water surface elevation being stored or routed, they are also known as crest gates.

In the case of flood bypass systems, floodgates sometimes are also used to lower the water levels in a main river or canal channels by allowing more water to flow into a flood bypass or detention basin when the main river or canal is approaching a flood stage.

Figure: *Tokyo floodgates created to protect from typhoon surges*

Types

- Bulkhead gates are vertical walls with movable, or removable, sections. Movable sections can be lifted to allow water to pass underneath (as in a sluice gate) and over the top of the structure. Historically, these gates used stacked timbers known as stop logs or wooden panels known as flashboards to set the dam's crest height. Some floodgates known as coupures in large levee systems slide sideways to open for various traffic. Bulkhead gates can also be made of other materials and used as a single bulkhead unit. Mitre gates are used in ship locks and usually close at an 18° angle to approximate an arch.
- Hinged crest gates, are wall sections that rotate from vertical to horizontal, thereby varying the height of the dam. They are generally controlled with hydraulic power, although some are passive and are powered by the water being impounded.

  Variations;

  - flap gate
  - fish-belly flap gates
  - Bascule gates
  - Pelican gates.

- Radial gates are rotary gates consisting of cylindrical sections. They may rotate vertically or horizontally. Tainter gates are a vertical design that rotates up to allow water to pass underneath. Low friction trunnion bearings, along with a face shape that balances hydrostatic forces, allow this design to close under its own weight as a safety feature.
- Drum gates are hollow gate sections that float on water. They are pinned to rotate up or down. Water is allowed into or out of the flotation chamber to adjust the dam's crest height.
- Roller gates are large cylinders that move in an angled slot. They are hoisted with a chain and have a cogged design that interfaces with their slot.
- Clamshell gates have an external clamshell leaf design.
- Fusegates are an innovative spillway control technology, which consists of free standing blocks (the Fusegates) set side by side on a flattened spillway sill. The Fusegate blocks act as a fixed weir most of the time and operate independently without any remote control or energy source only in case of excessive flood conditions. The System is developed and patented by Hydroplus from Paris, France. It has been installed on more than 50 dams around the world with sizes ranging from 1m to more than 9 m in height. Fusegate are typically used to increase the storage capacity of existing dams or to maximize the discharge potential of undersized spillways.

Valves

Valves used in floodgate applications have a variety of design requirements and are usually located at the base of dams. Often, the most important requirement (besides regulating flow) is energy dissipation. Since water is very heavy, it exits the base of a dam with the enormous force of water pushing from above. Unless this energy is dissipated, the flow can erode nearby rock and soil and damage structures.

Other design requirements include taking into account pressure head operation, the flow rate, whether the valve operates above or below water, and the regulation of precision and cost.

- Fixed cone valves, also known as Howell-Bunger valves, are designed to dissipate the energy from a water flow during reservoir discharge. They are a round pipe section with an adjustable sleeve gate and cone at the discharge end. Flow is varied by moving the sleeve away or towards its cone seat. The design allows high pressure water from the base of a dam to be released without causing erosion to the surrounding environment. Fixed cone valves are able to handle heads up to 300 m.

- Hollow jet valves are a type of needle valve used for floodgate discharge. A cone and seat are inside a pipe. Water flows through an annular gap between the pipe and cone when it is moved downstream, away from the seat. Ribs support the bulb assembly and supply air for water jet stabilisation.
- Ring jet valves are similar to fixed cone valves, but have an integral collar that discharges water in a narrow stream. They are suitable for heads up to 50 m.
- Jet flow gate, similar to a gate valve but with a conical restriction prior to the gate leaf that focuses the water into a jet. They were developed in the 1940's by the United States Bureau of Reclamation to allow fine control of discharge flow without the cavitation seen in regular gate valves. Jet flow gates are able to handle heads up to 150 m.

Physics

In order to do a simple calculation of the force on a rectangular flood gate one can use the following equation:

$$F = pA$$

where:

F = force measured in the SI units kg·m·s^{-2} which is called the newton (N)

p = pressure $= \rho g h$ measured in N/m^2, which is called the pascal (Pa)

where:

- ρ (rho) is the density of fresh water (1000 kg/m^3);
- g is the acceleration due to gravity on Earth (9.8 m/s^2);
- h is the height of the water column in metres.

A = area = rectangle : length × height measured in m^2

where:

length = the horizontal length of a rectangular floodgate measured in metres

height = the height of a non-submerged flood gate from the bottom of the water column to the water surface measured in metres

If the rectangular flood gate is submerged below the surface the same equation can be used but only the height from the water surface to the middle of the gate must be used to calculate the force on the flood gate.

Beach Nourishment

Beach nourishment— also referred to as beach replenishment— describes a process by which sediment (usually sand) lost through longshore drift or erosion is replaced from sources outside of the eroding beach. A wider beach can reduce storm damage to coastal structures by dissipating energy across the surf zone, protecting upland structures and infrastructure from storm surges, tsunamis and unusually high tides. Beach nourishment is typically part of a larger coastal defence scheme. Nourishment is typically a repetitive process, since nourishment does not remove the physical forces that cause erosion; it simply mitigates their effects. A benign environment increases the interval between nourishment projects, reducing costs. Conversely, high erosion rates may render nourishment financially impractical. In many coastal areas, the economic impacts of a wide beach can be substantial. The 10 miles (16 km), long shoreline fronting Miami Beach, Florida was replenished over the period 1976–1981. The project cost approximately $64,000,000 and revitalised the area's economy. Prior to nourishment, in many places the beach was too narrow to walk along, especially during high tide.

The first nourishment project in the U.S. was at Coney Island, New York in 1922-23 and is now a common shore protection measure utilised by public and private entities.

Nourishment is one of three commonly accepted methods for protecting shorelines. The structural alternative involves constructing a seawall, revetment, groin or breakwater. Alternatively, with "managed retreat" the shoreline is left to erode, while relocating buildings and infrastructure further inland. Nourishment gained popularity because it preserved beach resources and avoided the negative effects of hard structures. Instead, nourishment creates a "soft" (i.e., non-permanent) structure by creating a larger sand reservoir, pushing the shoreline seaward.

History

The first nourishment project in the U.S. was constructed at Coney Island, New York in 1922-1923.

Causes of Erosion

Beaches can erode both naturally and due to impact of humans.

Erosion is a natural response to storm activity. During storms, sand from the visible beach submerges to form storm bars that protect the beach. Submersion is only part of the cycle. During calm weather smaller waves return sand from the storm bar to the visible beach surface in a process called accretion. The term *erosion* conjures visions of environmental damage so the term submersion often replaces it in describing a healthy sandy beach. Continental drift erodes coastlines naturally. Ocean currents can change, disrupting the submersion/accretion cycle.

Some beaches do not have enough sand available to coastal processes to respond naturally to storms. When there is not enough sand left available on a beach, then there is no recovery of the beach following storms.

Many areas of high erosion are due to human activities. Reasons can include seawalls locking up sand dunes, coastal structures like ports and harbours that prevent longshore drift, dams and other river management structures. These activities interfere with the natural sediment flows either through dam construction (thereby reducing riverine sediment sources) or construction of littoral barriers such as jetties, or by deepening of inlets; thus preventing longshore transport of sediment across these channels.

Visible and Submerged Sand

The distinction between total sand in a beach and the proportion of the sand above the waterline (submersion fraction) critically impacts beach nourishment. Two beaches with the same amount of visible sand may look much different under water. An eroded beach with substantial submerged sand surrounding it may recover without nourishment. Nourishing a beach that has little submerged sand requires addressing the reason that the submerged sand is missing. Otherwise, the same forces that stripped the submerged sand once are likely to do so again. The amount of submerged sand eroded is typically much greater than the amount of missing sand on shore. Replacing only the visible sand is insufficient without replacing the sand that once supported the accretion part of the process is insufficient. In that circumstance, the beach is unstable and the visible sand quickly erodes. If human activity is a major cause of the erosion, mitigating that activity may be more cost effective over both short and long term periods than nourishment.

Requirements for Effective Nourishment

Sediment texture (grain size and sorting) is critical for success. Sand fill must be compatible with native beach sand. In some cases, beaches have been nourished using a finer sand than the original. Thermolumine scence monitoring reveals that storms can erode such beaches far more quickly than the natural beach. This was observed at the Waikiki nourishment project in Hawaii.

Profile Nourishment

Beach Profile Nourishment describes programs that nourish the full beach profile. In this instance, "profile" means the slope of the uneroded beach from above the water to well out to sea, not just the visible portion. The Gold Coast profile nourishment program placed 75% of its total sand volume below low water level. Some coastal authorities *overnourish* the below water beach (aka "nearshore nourishment") so that over time the natural beach increases in size. These approaches do not permanently protect beaches eroded by human activity. Doing so still requires mitigating that activity.

Replenishment Material

The selection of suitable material for a particular project depends upon the design needs, environmental factors, transport costs considering both short and long-term implications.

The most important material characteristic is the sediment's grain size, which must closely match the native material. Excess silt and clay fraction (mud) versus the natural turbidity in the nourishment area disqualifies some materials. Projects that did not match grain sizes performed relatively poorly. Nourishment sand that is only slightly smaller than native sand can result in significantly narrower equilibrated dry beach widths compared to sand the same size as (or larger than) native sand. Evaluating material fit requires a sand survey that usually includes geophysical profiles and surface and core samples.

Nourishment Projects

The setting of a beach nourishment project is key to design and potential performance. Possible settings include a long straight beach, an inlet that may be either natural or modified, and a pocket beach. Rocky or seawalled shorelines, that otherwise have no sediment, present unique problems.

Cancun, Mexico

Federal and state governments in Mexico have invested about \$71 million dollars (\$957 million pesos) throughout the state of Quintana Roo in restoring the beaches along Cancun, Playa del Carmen, and Cozumel.

Hurricane Wilma hit the beaches of Cancun and the Riviera Maya in 2005. The initial nourishment project was unsuccessful, leading to a second round that began in September 2009, and was scheduled to complete in early 2010. The project designers and the government committed to invest in beach maintenance to address future erosion. Project designers considered factors such as the time of year and sand characteristics such as density. Restoration in Cancun was expected to deliver 1.3 billion US gallons (4,900,000 m^3) of sand to replenish 450 metres (1,480 ft) of coastline. This time, the beach is promised to last at least 10 years.

Northern Gold Coast, Queensland, Australia

Gold Coast beaches in Queensland, Australia have experienced periods of severe erosion. In 1967 a series of 11 cyclones removed most of the sand from Gold Coast beaches. The Government of Queensland engaged engineers from Delft University in the Netherlands to advise them. The 1971 Delft Report outlined a series of works for Gold Coast Beaches, including beach nourishment and an artificial reef. By 2005 most of the recommendations had been implemented.

The Northern Gold Coast Beach Protection Strategy (NGCBPS) was a Aus \$10 million dollar investment. NGCBPS was developed between 1992 and 1999 and the works were completed between 1999 and 2003. The project included dredging 3,500,000 cubic metres (4,600,000 cu yd) of compatible sand from the Gold Coast Broadwater and delivering it through a pipeline to nourish 5 kilometres (3.1 mi) between of beach between Surfers Paradise and Main Beach. The new sand was stabilised by an artificial reef constructed at Narrowneck out of huge geotextile sand bags. The new reef was designed to improve wave conditions for surfing. A key monitoring program for the NGCBPS is the ARGUS coastal camera system operated by the University of New South Wales. The cost/benefit ratio for NGCBPS was conservatively estimated at 75:1 for a AUS \$10 million investment into beach replenishment. The benefits were estimated from a model of lost visitor nights in hotels following previous erosion events. NGCBPS so improved beach health that recovery following minor and moderate storms occurred within weeks. Additional unquantified benefits included lifestyle benefits for residents, additional public open space and improved fishing, diving and surfing conditions.

Netherlands

More than one-quarter of the Netherlands is below sea level and about 80% of the coast consists of sand dune or beach. The shoreline is closely monitored by yearly recording of the cross section at points 250 metres (820 ft) apart, to ensure adequate protection. Where long-term erosion is identified, beach nourishment using high-capacity suction dredgers is deployed.

Hawaii

Waikiki: Hawaii planned to replenish Waikiki beach in 2010. Budgeted at $2.5 million, the project covered 1,700 feet (520 m) in an attempt to return the beach to its 1985 width. Prior opponents supported this project, because the sand was to come from nearby shoals, reopening a blocked channel and leaving the overall local sand volume unchanged, while closely matching the "new" sand to existing materials. The project planned to apply up to 24,000 cubic yards (18,000 m^3) of sand from deposits located 1,500 to 3,000 feet (460 to 910 m) offshore at a depth of 10 to 20 feet (3.0 to 6.1 m). The project was larger than the prior recycling effort in 2006-07, which moved 10,000 cubic yards (7,600 m^3).

Maui

Maui, Hawaii illustrated the complexities of even small-scale nourishment projects. A project at Sugar Cove transported upland sand to the beach. The sand allegedly was finer than the original sand and contained excess silt that enveloped coral heads, smothering the coral and killing small animals that lived in and around it. As in other projects, onshore sand availability was limited, forcing consideration of more expensive offshore sources.

A second project, along Stable Road, that attempted to slow rather than halt erosion, was stopped halfway toward its goal of adding 10,000 cubic yards (7,600 m^3) of sand. The beaches had been retreating at a "comparatively fast rate" for half a century. The restoration was complicated by the presence of old seawalls, groins, piles of rocks and other structures. 7This project used sand-filled Geotube groins placed that were originally to remain in place for up to 3 years. The pipeline was anchored by concrete blocks attached by fibre straps. A video showed the blocks bouncing off the coral in the current, killing whatever they touched. In places the straps broke, allowing the pipe to move across the reef, "planing it down". Bad weather exacerbated the damaging movement and led to project shutdown. The smooth, cylindrical Geotubes could be difficult to climb over before they were covered by sand.

Supporters claimed that 2010's seasonal summer erosion was less than in prior years, although the beach was narrower after the restoration ended than in 2008. Authorities were studying whether to require the project to remove the groins immediately. Potential alternatives to Geotubes for moving sand included floating dredges and/or trucking in sand dredged offshore. A final consideration was sea level rise and that Maui was sinking under its own weight. Both islands surround massive mountains (Haleakala, Mauna Loa, and Mauna Kea) and were expanding a giant dimple in the ocean floor, some 30,000 feet (9,100 m) below the mountain summits.

Alternatives/Complements to Nourishment

Nourishment is not the only technique used to address eroding beaches. Others can be used singly or in combination with nourishment, driven by economic, environmental, and political considerations. Human activities such as dam construction can interfere with natural sediment flows (thereby reducing riverine sediment sources.) Construction of littoral barriers such as jetties and deepening of inlets can prevent longshore sediment transport.

Structural

The structural approach attempts to prevent erosion. Armouring involves building revetments, seawalls, detached breakwaters, groins, etc. Structures that run parallel to the shore (seawalls or revetments) prevent erosion. While this protects structures, it doesn't not protect the beach which is outside the wall. The beach disappears over a period that ranges from months to decades. Groynes and breakwaters that run perpendicular to the shore protect the shore from erosion. Filling a breakwater with imported sand can stop the breakwater from trapping sand from the littoral stream (the ocean running along the shore.) Otherwise the breakwater may deprive downstream beaches of sand and accelerate erosion there. Armouring may restrict beach/ ocean access, enhance erosion of adjacent shorelines, and requires long-term maintenance.

Managed Retreat

A second option is to move structures and infrastructure inland as the shoreline erodes. Retreat is more often chosen in areas of rapid erosion and in the presence of little or obsolete development.

Recruitment

Appropriately constructed and sited fences can capture blowing sand, building/restoring sand dunes, and progressively protecting the beach from the wind, and the shore from blowing sand.

Pressure Equalising Modules

All beaches lose and gain sand depending on the tides, precipitation, wind, waves etc. Wet beaches tend to lose sand. Waves infiltrate dry beaches easily and deposit sediment on the beach. Generally a beach is wet during falling tide, because the sea sinks faster than the beach drains. As a result most erosion happens during falling tide. PEMs allow the beach to drain more effectively during falling tide. Fewer hours of wet beach translate to less erosion. Permeable PEM tubes inserted vertically into the foreshore connect the different layers of groundwater. The groundwater enters the PEM tube allowing gravity to conduct it to a coarser sand layer, where it can drain more quickly. The PEM modules are placed in a row from the dune to the mean low waterline. Distance between rows is typically 300 feet (91 m) but this is project-specific. PEM systems come in different sizes depending. Modules connect layers with varying hydraulic conductivity. Air/water can enter and pressure equalize. PEMs are minimially invasive, typically covering approximately 0.0005% of the beach. The effects are local, and bring no known harm to flora or fauna, including nesting sea turtles. The tubes are below the beach surface, implying no visible impact. Installation takes little time. Costs are low, because PEMs require few materials, and no energy-intensive dredging. PEM installations have been successfully installed on beaches in Denmark, Sweden, Malaysia, and Florida, USA.

Measuring Project Impact

Nourishment projects usually involve physical, environmental and economic objectives. Typical physical measures include dry beach width, remaining post-storm sand volume, post-storm damage avoidance assessments and aqueous sand volume. Environmental measures include marine life distribution, habitat and population counts. Economic impacts include recreation, tourism, flood and "disaster" prevention. Techniques for incorporating nourishment projects into flood insurance costs and disaster assistance remain controversial. The ability to predict the performance of a beach nourishment project is best for a project constructed on a long, straight shoreline without the complications of inlets or engineered structures. In addition, predictability is better for overall performance, e.g., average shoreline change, rather than shoreline change at a specific location. Nourishment can affect eligibility in the National Flood Insurance Program and federal disaster assistance.

Environmental Issues

Beach nourishment has significant impacts on local ecosystems. Nourishment may cause direct mortality to sessile organisms in the target area by burying them in the new sand. Seafloor habitat in both source and target areas are disrupted, e.g., when sand is deposited on coral reefs or when deposited sand hardens. Imported sand may differ in character (chemical makeup, grain size, non-native species) from that of the target environment. Reduced light availability, affecting nearby reefs and submerged aquatic vegetation. Imported sand may contain material toxic to local species. Removing material from near-shore environments may destabilize the shoreline, in part by steepening its submerged slope. Related attempts to reduce future erosion may provide a false sense of security that increases development pressure.

Sea Turtles

Newly-deposited sand can harden and complicate nest-digging for turtles. However, nourishment can provide more/better habitat for them, as well as for sea birds and beach flora. Florida addressed the concern that dredge pipes would suck turtles into the pumps by adding a special grill to the dredge pipes that ensured that turtles were not sucked in with the sand.

Sand Dune Stabilisation

Sand dunes are common features of shoreline and desert environments. Dunes provide habitat for highly specialised plants and animals, including rare and endangered species. They can protect beaches from erosion and recruit sand to eroded beaches. Dunes are threatened by human activity, both intentional and unintentional. Countries such as the United States, Australia, Canada, New Zealand, the United Kingdom, and Netherlands, operate significant dune protection programs. Stabilising dunes involves multiple actions. Planting vegetation reduces the impact of wind and water. Fences catch sand and other material. Footpaths protect dunes from damage from foot traffic.

Vegetation

The location on the dune limits the types of plant that can thrive there. Beach dunes consist of the *foredune*, the angled side which faces the ocean, the *sand plain* at the top of the dune, which may or may not be present, and the *backdune*, the angled side that faces away from the ocean.

Foredune Flora

Plants that thrive on the foredune must be tolerant to salt spray, strong winds, and burial by blowing/accumulating sand. Typical vegetation includes Ammophila arenaria, Honckenya peploides, Cakile maritima, and Spartina coarctata.

Backdune Flora

Plants which thrive on the broad dune plain and backdune grow together into dense patches termed *dune mats* that hold the dune together. Vegetation typical of the plain and backdune include Hudsonia tomentosa, Spartina patens, Iva imbricata, and Eregeron glaucus. Introduced species can outcompete native plants and disrupt animal life, making them formally "invasive species".

Shrub Stage

The above species are herbaceous plants. After they have rooted and developed fully, a second stage, the "shrub stage", can begin. During this phase, larger plants with deeper root systems can be planted.

Examples are Empetrum nigrum, Ilex vomitoria, and Vaccinium ovatum. The shrub stage is usually the final phase and may last for short or long periods of time depending on microclimatic conditions such as distance from the shoreline, availability of ground water, or salt spray effects..

Coastal Management

A single beach may be divided into segments with different owners. Gaining agreement among all owners complicates the process of stabilising the dunes. Without agreement, some parts of the dune may go unplanted, while others host visually appealing plants that do little to stabilize the dune. Inconsistent vegetation, known as a fragmented or decoupled gradient, can create weak points in the dune that limit its effectiveness against floods and even its continued existence.

Publicly owned beaches, found in U.S. states such as California and Hawaii and in other jurisdictions, present the opportunity to systematically manage—or mismanage—beaches and their accompanying dunes.

Coastal Sand Dune Management

Figure : *Spencer Park restoration*

Coastal sand dunes fringe thousands of Kilometres of coastline around the world. They are made up of continuous, hummocky hills of sand that are held together by specially adapted sand dune vegetation. See Figure 1. These plants have adaptations that allow them to survive in sand dune habitats, these include adaptations to a moving substrate which covers and uncovers its roots, adaptation to desiccation, to strong winds carrying salt and sand and to wildly changing temperatures.

Coastal sand dunes formed during a time of low sea level where sand and sediments on the foreshore became exposed and in conjunction with sufficient winds, this sand was transported up the beach face via a process known as saltation.

Once sufficient wind (>15 km/hr) mobilises sand particles they become trapped in vegetation or drift wood at the back of the beach, sand then continues to accumulate among the vegetation and over time slowly builds up the dunes we see today. Sand dunes serve an important purpose by protecting inland areas from coastal water intrusion. They are able to absorb the impact and protect inland areas from high energy storms and act as a resilient barrier to the destructive forces of wind and waves.

Human Impacts

It is generally observed that coastal dunes are highly sensitive to any number of human activities. It is also understood that coastal dunes need to have limited interaction with humans in order for their protection and survival. Around the world coastal dunes are eroding and under pressure due to increased sediment loss because of increased coastal human activities. For example, in the last 30 years, mainly because of tourism, nearly 75% of the Mediterranean's coastal dunes have been damaged or destroyed. Human impacts can expand from minor to highly significant adverse effects. Minor effects include pedestrian trampling, constructing footpaths, and off road vehicle tracks, to the highly more significant which includes the construction of roads, car parks, houses, and the use of sand dunes for ground water extraction, the conversion of sand dunes to land for farming, forestry, waste disposal and sand dune mining.

New Zealand is a country with a very long coast line that consists of around 1090 km of sand dunes. From the early 1900's to the early 2000's the area of sand dunes has declined by about 70% from 129,000 ha to 39,000 ha. This decline has been mainly due to erosion caused by the depletion of vegetation cover in order to convert sand dunes to land for forestry, farming and urban areas. Loss in vegetation cover exposes sand to the action of wind and leaves sand dunes vulnerable to increased erosion as it is vegetation that binds sand together. The loss of vegetation causes sand to move inland which is a process known as sand encroachment, this process can having damaging effects on inland properties and leaves properties vulnerable to future coastal hazards for example storm surge.

As well as human activities, natural processes operate within coastal dune environments, these processes include events such as storms, floods, sea level change, and sediment supply changes. Coastal dune systems are mainly controlled by a principle natural process called the sediment supply and transport system. This system consists of an offshore zone (sediment bank), transit zone (beach and active foredune) and a resting zone (stable sand dunes). The sediment in this system is supplied by long shore drift predominantly from rivers and from the erosion of cliffs, headlands, other dune systems and when sediment supply is low sand dunes are subject to erosion. Humans affect sediment supply in a number of ways including the damming of rivers, dredging in coastal water and the construction of harbours, all of which depletes sediment supply. They also disrupt the long shore transport of sediment through the construction of piers, marinas or groynes, breakwaters and other hard structures. Excavating, building roads, houses and walkways on coastal dunes or beaches also causes sand dune erosion by disrupting natural sediment processes. Effective management of coastal dune systems must consider natural processes as a normal part of the natural environment as well as controlling human activities and seeking to alter natural processes should be kept at a minimum in order for restoration to be successful.

Integrated Management

Sand dunes are an important part of the coast as they are land's first defence against the forces of the ocean. Because of this, management of sand dunes will focus on the natural authenticity and values of the dunes themselves including flora, fauna, and the ecosystem as a whole. The concept of integrated management can be incorporated into management of coastal sand dunes. This concept is made up of three main aspects:

- Restore and conserve natural processes
- Create conditions for future natural processes
- Develop types of land use that agree with the above.

These concepts seek to incorporate natural processes and human activities within coastal dunes systems. Sand dune management issues are related to the nature of sediment, which has no cohesiveness so the influence of wind and sediment supply is an important factor in determining erosion and deposition of dry sand grains.

Sand dune management consists of restoring natural dune systems and processes with the implementation of "soft" technique which includes the cessation of any activity that adversely effects the sediment supply of the dunes, planting vegetation, fencing off sensitive areas, creating dune walkways, and providing information to the public regarding the importance of maintaining stable sand dunes. The concept of "soft" techniques is in contrast to the widely used "hard" techniques which consist of protecting the coast through the construction of concrete sea walls, revetments, gabions, groynes and offshore break waters. The problem with using these "hard" engineering methods is that they only provide temporary protection against ocean hazards and each "hard" method produces an "end effect" that causes erosion problems to adjacent coastal areas. This is why the implementation of "soft" techniques in dune restoration should be used more widely in coastal protection as they provide natural, long term protection with no adverse effect to adjacent coastal areas. The use of soft techniques can be seen in many coastal protection projects around the world including many in New Zealand.

Restoration and Protection

One of the first things a coastal manager must look for when establishing sand dune management is to ascertain the main cause of sediment loss and from there, make decisions on how to minimise the cause and how to restore sand dunes appropriately. In New Zealand the main cause of dune erosion has been the loss of sand vegetation and this means that sand dune management consists primarily of stabilising dune systems through the "soft" technique of re planting and protecting native dune vegetation. It is advised that dunes should not be restored with the planting of exotic species and managers should aim to only use species indigenous to an area, originally New Zealand used exotic, fast growing dune vegetation called Marram Grass (Ammophila arenaria) to stabilize eroding sand dunes. It was later discovered that this exotic species created very high, steep sand dunes that were increasing susceptible to New Zealand's high storm energy climate, it also became apparent that the native New Zealand dune vegetation were out competed by this introduced species which caused issues biologically. At present there is a large push to eradicate Marram grass as it is now classed as a pest species in New Zealand and at present coastal managers encourage the planting of the two indigenous species, Pingao (Desmoschoenus spiralis) and Spinifex (Spinifex sericeus) as these two species create low angle, strong, stable dunes.

The management of dune restoration projects includes planting vegetation successfully. Merely scattering and leaving seeds over sandy areas is perceived as a potential waste as strong winds can blow seeds away. Planting of nurseries should be conducted in the hollows of any present sand hills or in depressions of sandy areas. Seeds should be planted in moist conditions and should be lightly covered by a layer of sand and large mature grass individuals should be planted in deep holes with the tops of the plants cut short so as to reduce wind resistance. Planting of sand dune vegetation alone does not entirely restore or protect sand dunes from human impacts. Seedlings need to be protected from human trampling which is done mainly by fencing off newly planted areas. Fencing also acts as another way of trapping and accumulating sand among the dunes and can also act as protection against strong winds.

The Public

Because sand dunes and beaches are extremely popular for outdoor recreation, the success of restoration and dune management depends a great deal on the information available and understanding of the public. This is important because not all aspects of sand dune management are obvious or important to the public if they do not understand why management is necessary e.g stabilisation. There are many ways in which information can be provided to the public including the erection of panels at entrances to dune walkways, leaflets, brochures, maps and council workers who can talk to the public about the importance of protecting sand dunes. Figure five shows a photo of a panel put up at Spencer Park by the Christchurch City Council indicating that people should not walk on the area, as it is an area for sand dune restoration. It needs to be reiterated that sand dune restoration and protection is in the public's best interest as the loss of sediment on any coast line leads to the erosion of beaches, which in turn leads to a loss of recreational value of many coast lines. The public needs to be aware of the role that sand dunes play in coastal protection because without public support coastal sand dune management will be virtually impossible.

Beach Drainage

Beach drainage or beach face dewatering lowers the water table locally beneath the beach face. This causes accretion of sand above the drainage system. Grant (1946) – the elevation of the beach water table had an important bearing on deposition and erosion across the foreshore.

A high water table coincided with periods of accelerated beach erosion, and conversely, a low water table coincided with pronounced aggradation of the foreshore A lower water table (unsaturated beach face) facilitates deposition by reducing flow velocities during backwash and prolonging laminar flow. In contrast, a high water table results in condition favouring beach erosion. With the beach in a saturated state, Grant proposed that backwash velocity is accelerated by the addition of ground water seepage out of the beach within the effluent zone.

Turner and Leatherman (1997) moving from the origins and development of the dewatering concept to field and laboratory studies available at the time of writing concluded that there was too little evidence for being convinced that the systems had a positive effect. None of the case studies provide full scientific evidence of indisputable positive results regarding beach stabilisation although in some cases an overall positive performance was reported. In many cases no adequate long-term monitoring was undertaken at a frequency high enough to discriminate the response to high energy erosive events.

A useful side effect of the system is that the collected seawater is very pure because of the sand filtration effect. It may be discharged back to sea but can also be used to oxygenate stagnant inland lagoons /marinas or used as feed for heat pumps, desalination plants, land based aquaculture, aquariums or seawater swimming pools.

Beach drainage systems have been installed in many locations around the world to halt and reverse erosion trends in sand beaches. Twenty four beach drainage systems have been installed since 1981 in Denmark, USA, UK, Japan, Spain, Sweden, France, Italy and Malaysia.

Costs

The costs of installation and operation per metre of shoreline protection will vary due to:

- system length (non-linear cost elements)
- pump flow rates (sand permeability, power costs)
- soil conditions (presence of rock or impermeable strata)
- discharge arrangement /filtered seawater utilisation
- drainage design, materials selection & installation methods
- geographical considerations (location logistics)
- regional economic considerations (local capabilities /costs)
- study requirements /consent process.

The costs associated with a beach drainage system are generally considerably lower than hard engineered structures. They also compare very favourably with beach nourishment projects, particularly when long-term project economics are considered (nourishment projects often have a limited life or a program of re-nourishment).

Monitoring Coastal Zones

Coastal zone managers are faced with difficult and complex choices about how best to reduce property damage in the shorelines. One of the problems they face is error and uncertainty in the information available to them on the processes that cause erosion of beaches. Video-based monitoring lets collect data continuously at low cost and produce analyses of shoreline processes over a wide range of averaging intervals.

Event Warning Systems

Event warning systems, such as tsunami warnings and storm surge warnings, can be used to minimize the human impact of catastrophic events that cause coastal erosion. Storm surge warnings can also be used to determine when to close floodgates to reduce the physical impact of such events. Wireless sensor networks can be deployed quickly to set up a coastal erosion monitoring system, and scaled accordingly.

Shoreline Mapping

Defining the shoreline is a difficult task due to the dynamic nature of the coast and the intended application of the shoreline (Graham et al. 2003; Boak & Turner 2005). Given this idea the shoreline must therefore be considered in a temporal sense whereby the scale is dependent on the context of the investigation (Boak & Turner 2005).

The following definition of the coast and shoreline is most commonly employed for the purposes of shoreline mapping. The coast comprises the interface between land and sea, and the shoreline is represented by the margin between the two (Woodroffe, 2002). Due to the dynamic nature of the shoreline coastal investigators adopt the use of shoreline indicators to represent the true shoreline position (Boak & Turner 2005).

Shoreline Indicator

The choice of shoreline indictor is a primary consideration in shoreline mapping. According to Leatherman (2003) it is important that indicators are easily identified in the field and on aerial photography. Shoreline indicators may be physical beach morphological features such as the berm crest, scarp edge, vegetation line, dune toe, dune crest and cliff or the bluff crest and toe. Alternatively, non-morphological features may also be used. These indicators are based on water level including the high water line, mean high water line, wet/dry boundary, and the physical water line (Pajak & Leatherman 2000). Figure 1 provides a sketch of the spatial relationship between many of the commonly used shoreline indicators. The high water line (HWL), defined as the wet/dry line is the most commonly used shoreline indicator because it is visible in the field, and can be interpreted on both colour and grey scale aerial photographs (Leatherman, 2003; Crowell et al. 1991). The HWL represents the landward extent of the most recent high tide and is characterised by a change in sand colour due to repeated, periodic inundation by high tides. The HWL is portrayed on aerial photographs by the most landward change in colour or grey tone (Boak & Turner 2005).

Importance and Application

The location of the shoreline and its changing position over time is of fundamental importance to coastal scientists, engineers and managers (Boak & Turner 2005; Pajack & Leatherman 2002). Present day shoreline monitoring campaigns provide information about historic shoreline location and movement, and about predictions of future change (Appeaning Addo et al. 2008). More specifically the position of the shoreline in the past, at present and where it is predicted to be in the future is useful for in the design of coastal protection, to calibrate and verify numerical modells to assess sea level rise, map hazard zones and formulate policies to regulate coastal development. Accurate and consistent delineation of the shoreline is integral to all of these tasks. The location of the shoreline also provides information regarding shoreline reorientation adjacent to structures, beach width, volume and rates of historical change (Boak & Turner 2005; Pajack & Leatherman 2002).

Data Sources

A variety of data sources are available for examining shoreline position however, the availability of historical data is limited at many coastal sites and so the choice of data source is largely limited to what is available for the site at a given time (Boak & Turner 2005). Shoreline mapping techniques applied to data sources have moved towards automation in association with technological advances and the need to reduce uncertainty. Although these changes have resulted in improvement in coastal data processing and storage capabilities, the frequent change in technology has prevented the emergence of one standard method of shoreline mapping. This has occurred because each data source and associated method have their own unique capabilities and shortcomings (Moore 2000). A number of the data sources used for shoreline mapping and their associated advantages and disadvantages.

Historical Maps

In the event that a study requires the shoreline position to be mapped before the development of aerial photographs, or if the location has poor photograph coverage it is necessary to employ historical maps in order to detail shoreline position (Moore 2000). The main advantage and reason for using historical maps is that they are able to provide a historic record that is not available from other data sources. Many potential errors however are associated with historical coastal maps and charts. Such errors may be associated with scale, datum changes, distortions from uneven shrinkage, stretching, creases, tears and folds, different surveying standards, different publication standards, and projection errors (Boak & Turner 2005). The severity of these errors depends on the accuracy standards met by each map and the physical changes that have occurred since the publication of the map (Anders & Byrnes 1991). The oldest reliable source of shoreline data in the United States dates back to the early to mid 19th century and is the U.S Coast and Geodetic Survey/National Ocean Service T-sheets (Morton 1991). In the United Kingdom, many maps and charts were deemed to be inaccurate until around 1750. The founding of the Ordnance Survey in 1791 has since improved the accuracy of the mapping.

Aerial Photographs

Aerial photographs have been used since the 1920's to provide topographical information about an area. They are therefore a good database for compilation of historical shoreline change maps. Aerial photographs are the most commonly used data source in shoreline mapping because many coastal areas have extensive aerial photo coverage therefore providing a valuable record of shoreline position (Moore 2000). In general, aerial photographs provide good spatial coverage of the coast however temporal coverage is very much site specific depending on the flight path of the aeroplane. A second disadvantage associated with aerial photography is that the interpretation of the shoreline position is subjective given the dynamic nature of the coastal environment. This combined with various distortions inherent in aerial photographs can lead to significant error levels (Moore 2000).

Object Space Displacements

Conditions outside of the camera can cause objects in an image to be displaced from their true ground position. Such conditions may include ground relief, camera tilt and atmospheric refraction.

Relief displacement is prominent when photographing a variety of elevations. This situation causes objects above ground level to be displaced outward from the centre of the photograph and objects below ground level to be displaced toward the centre of the image (Figure 2). The severity of the displacement is affected negatively with decreases in flight altitude and as radial distance from the centre of the photograph increases. This distortion can be minimised by photographing numerous swaths and creating a mosaic of the images. This technique will create a focus for the centre of each photograph where distortion is minimised. It is important to note that this error is not common in shoreline mapping is the relief is fairly constant. It is however important to consider when mapping cliffs (Moore 2000).

Ideally aerial photographs are taken so the optical axis of the camera is perfectly perpendicular to the ground surface thereby creating a vertical photograph. Unfortunately this is not often the case and virtually all aerial photographs experience tilt whereby up to 3° is not uncommon (Camfield et al. 1996). In this situation the scale of the image will be larger on the upward side of the tilt axis and smaller on the downward side. Moore, (2000) notes that many coastal researchers have not realised the severity of this error and therefore do not consider it in their methods.

Radial Lens Distortion

Lens distortion varies as a function of radial distance from the isocentre of the photograph meaning that the centre of the image is relatively distortion free, but as the angle of view increases the distortion becomes more prominent. This is a significant source of error in earlier aerial photography but as technology has increased and camera lens have become more refined it has become less of an issue with later photographs. Such a distortion is impossible to correct for without knowing the make and model of the lens used to capture the image. However if overlapping images have been acquired one can digitize the centre portions of the aerial photographs (Crowell et al. 1991).

Delineation of the Shoreline

The dynamic nature of the coast has meant that accurate mapping of an instantaneous shoreline position has been associated with significant uncertainty. This uncertainty arises because at any given time the position of the shoreline is influenced by the short-term effect of the tide and a wide variety of long term effects such as relative sea-level rise and along shore littoral sediment movement. Not only does this affect the accuracy of computed historic shoreline position but also any predicted future positions (Appeaning Addo et al. 2008). As mentioned earlier the HWL is most commonly used as a shoreline indicator. This can usually be seen as a significant tonal change on aerial photographs. There are however many errors associated with using the wet/dry line as a proxy for the HWL and shoreline. The errors of largest concern are the short term migration of the wet/dry line, interpretation of the wet/dry line on a photograph and measurement of the interpreted line position (Leatherman 2003; Moore 2000). Systematic errors such as the migration of the wet/dry line may arise from tidal and seasonal changes.

Storm-induced erosion is another factor which may cause the wet/dry line to migrate landward. Field investigations have shown that these changes can be minimised by using only summertime data (Moore 2000; Leatherman 2003). Furthermore, the error bar can be significantly reduced by using the longest record of reliable data to calculate erosion rates (Leatherman 2003). Finally it is important to note that errors may arise due to the difficulty of measuring a single line on a photograph. For example where the pen line is 0.13 mm thick this translates to an error of ±2.6 m on a 1:20000 scale photograph.

Beach Profiling Surveys

Beach profiling surveys are typically repeated at regular intervals along the coast in order to measure short-term (daily to annual) variations in shoreline position and beach volume. (Smith & Zarillo 1990). Beach profiling is a very accurate source of information however measurements are generally subject to the limitations of conventional surveying techniques. Shoreline data derived from beach profiling is often spatially and temporally limited due to the high cost associated with such a labour intensive activity. Shorelines are generally derived by interpolating between a series of discrete beach profiles. It is important to note however that the distance between the profiles is usually quite large and so the accuracy of the interpolating becomes compromised. In contrast to aerial photographs, survey data is limited to smaller lengths of shoreline generally less than ten kilometres (Boak & Turner 2005). Beach profiling data is commonly available in from regional councils in New Zealand such as those compiled by the Hawkes Bay Regional Council.

Remote Sensing

Technological advancement over the last decade has led to the development of a range of airborne, satellite and land based remote sensing techniques (Smith & Zarillo 1990). Some of the remotely sensed data sources are listed below:

- Multispectral and hyperspectral imaging
- Microwave sensors

- Global positioning system (GPS)
- Airborne light detection and ranging technology (LIDAR).

Remote sensing techniques are attractive as they are cost effective, reduce manual error and remove the subjective approach of conventional field techniques (Maiti et al. 2009). Remote sensing is a relatively new concept and so extensive historical observations are unavailable. Given this idea, it is important that coastal morphology observations are quantified by coupling remotely sensed data with other sources of information detailing historic shoreline position from archived sources (Appeaning Addo et al. 2008).

Video Analysis

Video analysis provides quantitative, cost-effective, continuous and long-term monitoring beaches (Turner et al. 2004). The advancement of coastal video systems over the past 15 years has resulted in the extraction of large amounts of geophysical data from images. Such data includes that about coastal morphology, surface currents and wave parametres. The main advantage of video analysis lies in the ability to reliably quantify these parametres with high resolution and coverage in both space and time. This in particular highlights their potential importance as an effective coastal monitoring system and an aid to coastal zone management (Van Koningsveld et al. 2007). Interesting case studies have been carried out using video analysis. Turner et al. (2004) used a video-based ARGUS coastal imaging system to monitor and quantify the regional-scale coastal response to sand nourishment and construction of the world-first Gold Coast artificial (surfing) reef in Australia. In addition, Smit et al. (2007) demonstrated the added value of high resolution video observations for making short-term predictions of near shore hydrodynamic and morphological processes, at temporal scales of metres to kilometres and days to seasons.

2

Construction Engineering Basics

Construction engineering is a professional discipline that deals with the designing, planning, construction, and management of infrastructures such as highways, bridges, airports, railroads, buildings, dams, and utilities. Construction Engineers are unique such that they are a cross between civil engineers and construction managers. Construction engineers learn the designing aspect much like civil engineers and construction site management functions much like construction managers.

Figure : *The Berlin Brandenburg Airport under construction. The facility starts full operation in 2012.*

The primary difference between a construction engineer and a construction manager is that the construction engineer has the ability to sit for the Professional Engineer license (PE) whereas a construction manager cannot. At the educational level, construction managers are not as focused on design work as they are on construction procedures, methods, and people management. Their primary concern is to deliver a project on time, with in budget, and of the desired quality.

The difference between a construction engineer and civil engineer is only at the educational level as both disciplines are able to sit for the PE exam giving them the same title of engineer. Civil engineering students concentrate more on the design work, gearing them toward a career as a design professional. This essentially requires them to take a multitude of design courses. Construction engineering students take design courses as well as construction management courses. This allows them to understand both the design functions as well as the building requirements needed to design and build today's infrastructures.

Work Activities

Depending on which career path the construction engineer has chosen to follow, an entry-level design engineer normally provide support to project managers and assist with creating conceptual designs, scopes, and cost estimates for the planning and construction of approved projects. It should be noted that a career in design work does require a professional engineer license (PE). Individuals who pursue this career path are strongly advised to sit for the Engineer In Training exam (EIT) while in college as it takes five years post graduate to obtain the PE license. Entry-level construction manager positions are typical called project engineers or assistant project engineers. They are responsible for preparing purchasing requisitions, processing change orders, preparing monthly budgeting reports, and handling meeting minutes. The construction management position does not necessarily require a PE license; however possessing one does make the individual more marketable, as the PE license allows the individual to sign off on temporary structure designs.

Abilities

Construction engineers are problem solvers, they help create infrastructure that best meets the unique demands of its environment. They must be able to understand infrastructure life cycles and have the perspective to solve technical challenges with clarity and imagination. Therefore individuals should have a strong understanding of math and science, but many other skills are required, including critical and analytical thinking, time management, people management and good communication skills.

Educational Requirements

Individuals looking to obtain a construction engineering degree must first ensure that the program is accredited by EAC or Technology Accreditation Commission (TAC) of the Accreditation Board for Engineering and Technology (ABET). ABET accreditation is assurance that a college or university program meets the quality standards established by the profession for which it prepares its students. In the US there are currently twenty-five programs that exist in the entire country so careful college consideration is advised.

A typical construction engineering curriculum is a mixture of engineering mechanics, engineering design, construction management and general science and mathematics. This usually leads to a Bachelor of Science degree. The B.S. degree along with some design or construction experience is sufficient for most entry level positions. Graduate schools may be an option for those who want to go further in depth of the construction and engineering subjects taught at the undergraduate level. In most cases construction engineering graduates look to either civil engineering, engineering management, or business administration as a possible graduate degree. For authority to approve any final designs of public projects (and most any project), a construction engineer must have a professional engineers (P.E.) license.

Job Prospects

Job prospects for construction engineers generally have a strong cyclical variation. For example, starting in 2008 - continuing until at least 2011 - job prospects have been poor due to the collapse of housing bubbles in many parts of the world. This sharply reduced demand for construction, forced construction professionals towards infrastructure construction and therefore increased the competition faced by established and new construction engineers. This increased competition, and a core reduction in quantity demand is in parallel with a possible shift in the demand for construction engineers due to the automation of many engineering tasks, overall resulting in reduced prospects for construction engineers. In early 2010 the United States construction industry had a 27% unemployment rate, this is nearly three times higher than the 9.7% national average unemployment rate. The construction unemployment rate (including tradesmen) is comparable to the United States 1933 unemployment rate - the lowest point of the Great Depression - of 25%.

Remuneration

The average salary for a civil engineer in the UK depends on the sector, and more specifically the level of experience of the individual. A 2010 survey of the remuneration and benefits of those occupying jobs in construction and the built environment industry showed that the average salary of a civil engineer in the UK is £29,582. The average salary varies depending on experience, for example the average annual salary for a civil engineer with between 3 and 6 years experience is £23,813. For those with between 14 and 20 years experience the average is £38,214.

Highway Engineering

***Figure** : Brazil's Rodovia Anchieta.*

Highway engineering is an engineering discipline branching from civil engineering which involves the design, construction and maintenance of Highway Systems. Highway Engineering become prominent towards the latter half of the 20th Century after World War 2. Standards of highway engineering are continuously being improved. Highway Engineers must take into account future traffic flows, design of highway intersections/interchanges, geometric alignment and design, highway pavement materials and design, structural design of pavement thickness, and pavement maintenance.

History of Highway Engineering

The beginning of road construction could be dated back to when the Romans were around. With the advancement of technology from carriages pulled by two horses to vehicles with power equivalent to 100 horses, road development had to follow suit. The construction of modern highways did not begin until the late 19th to early 20th century. The first research dedicated to highway engineering was initiated in the United Kingdom with the introduction of the Transport Research Laboratory (TRL), in 1930. In the USA, highway engineering became an important discipline with the passing of the Federal Aid Highway Act of 1944, which aimed to connect 90% of cities with a population of 50,000 or more. With constant stress from vehicles which grew larger as time passed, improvements to pavement were needed. With technology out of date, in 1958 the construction of the first motorway in Great Britain (the Preston bypass) played a major role in the development of new pavement technology.

Design policies standards used in the United States are typically based on publications of the American Association of State Highway and Transportation Officials as well as research promulgated by the Transportation Research Board, the Institute of Transportation Engineers, the Federal Highway Administration, and the Department of Transportation.

Bridge Engineering

Bridge is a structure which provides a passage to people, vehicles, railways or pipelines to cross various obstacles to travel. Engineers build bridges over obstacles such as lakes, rivers, canyons, and dangerous roads and railway tracks. Without bridges, people would need boats to cross waterways and would have to travel around canyons and ravines. Bridges range in length from a few metres to several kilometres. They are among the largest structures built by man. The demands on design and on materials are very high. A bridge must be strong enough to support its own weight as well as the weight of the people and vehicles that use it. The structure also must resist various natural occurrences, including earthquakes, strong winds, and changes in temperature. Most bridges have a concrete, steel, or wood framework and an asphalt or concrete roadway on which people and vehicles travel.

Basic Components of a Bridge

The majority of bridges are held up by at least two supports set in the ground. The distance between two adjacent supports is called a span of a bridge. The supports at each end of the bridge are called abutments, and the supports that stand between the abutments are called piers. The total length of the bridge is the distance between the abutments. Most short bridges are supported only by abutments and are known as single-span bridges. Bridges that have one or more piers in addition to the abutments are called multi-span bridges. Most long bridges are multi-span bridges.

In special cases, like in a pontoon bridge no piers or abutments are provided. It is supported by pontoons (flat-bottomed boats) or other portable floats. There are three main components of a typical bridge in general, namely:

- Superstructure
- Substructure
- Foundation.

The above three components have the following basic parts:

Superstructure :

a) The main constituent of any superstructure is the decking, which comprises of a slab, girder, trusses, etc.

b) The handrails, crash barriers, kerbs, etc.

The three basic types of bridge spans are shown below. Any of these spans may be constructed using beams, girders or trusses. Arch bridges are either simple or continuous (hinged). A cantilever bridge may also include a suspended span.

Substructure :

a) Abutments at the extreme ends of the bridge.

b) Piers at intermediate supports in case of multiple span bridges.

c) Bearings and pedestals for the decking.

Foundation :

a) These include foundations for both abutments and piers. Foundations may be of the type open, well, pile, etc.

Apart from the above, river training works and the approaches to a bridge also form a part of a bridge works.

Selection of Bridge Site

It may not be possible always to have a wide choice of sites for a bridge. This is particularly so in case of bridges in urban areas and flyovers. For rivers bridges in rural areas, usually a wider choice may be available.

The characteristics of an ideal site for a bridge across a river are:

- A straight reach of the river;
- Steady river flow without serious whirls and cross currents;
- A narrow channel with firm banks;
- Suitable high banks above high flood level on each side;
- Rock or other hard inerodible strata close to the river bed level;
- Economical approaches which should not be very high or long or liable to flank attacks of the river during floods; the approaches should be free from obstacles such as hills, frequent drainage crossing, sacred places, graveyards, or built up areas or troublesome land acquisition;
- Proximity to a direct alignment of the road to be connected;
- Absence of sharp curves in the approaches;
- Absence of expensive river training works; and
- Avoidance of excessive underwater construction.

For selecting a suitable site for a major bridge, the investigating engineer should make a reconnaissance survey for about one km on the upstream side and one km on the downstream side of the proposed bridge site and should journey along the road for about one km on either side of the road from the bridge site in order to form a general impression of the landscape and to decide on the type of structure best suited to the site. Care should be taken to investigate a number of probable alternative sites and then decide on the site which is likely to serve the needs of the bridge at the least cost. A brief description of the reasons for selection of a particular site should be furnished in the investigation report along with salient details of alternative sites investigated and rejected.

When the river to be crossed is a meandering river, the bridge should be located at a nodal point, i.e. the location where the river regime is constant serving as a fulcrum about which the river channels swing laterally. To the extent possible, it is desirable to align the bridge at right angles to the river, i.e. to provide a square crossing. Sometimes, a skew crossing which is included to the centre line of the river at an angle different from a right angle has to be provided in order to avoid costly land acquisition or sharp curves on the approaches. A skew bridge usually poses more difficulties in design, construction and maintenance. The location of the bridge in relation to the alignment of the approaches may be decided as below : (a) For bridges of total length less than 60 m, the alignment of the approach will govern; (b) For bridges of total length between 60 m and 300 m, both the proper alignment of the approaches and the requirements of a good bridge site should be considered together in ascertaining the appropriate site; and © For major bridge over 300 m length, the requirements of a good bridge site will govern the alignment.

Bridge Classification

The types of bridges vary in total length, the length of their spans, and the weight they can support. Before deciding which kind to build at a particular place, engineers determine the length of the structure and of each span. They also consider the maximum load the bridge will carry and the materials available to construct it. Considering various aspects, bridges may be classified in the following ways :

1. *Function or purpose to be performed :* Based on this bridges can be classified as an aqueduct over a river or a canal, a viaduct which may be road or railway over a valley, a pedestrian or a foot bridge, a highway bridge, a railway bridge, a road cum rail bridge, etc.
2. *Construction material used :* Under this heading bridges can be grouped as timber bridges, masonry bridges, steel bridges, reinforced cement concrete bridges, prestressed bridges, composite bridges, etc.
3. *Alignment of bridge :* Based on this bridges can be classified as a straight or a skew bridge.
4. *Location of bridge floor relative to the superstructure :* In accordance with the relative position of bridge floor as compared to the superstructure, bridges can be classified as deck, through, semi-through and suspended bridge.
5. *According to the type of superstructure :* Depending upon the type of superstructure, bridges can be grouped as slab, girder, truss, suspension, cable stayed, arch, drawbridges, etc.
6. *Nature of superstructure action :* Under its category bridges can be classified as portal frame bridges, truss bridges, balanced cantilever bridges and suspension bridges.
7. *Location of Highest Flood Level :* Particularly for a highway bridge, under this category bridges may be classified as a high level or non - submersible bridge and a submersible bridge

depending upon the formation level of the bridge relative to the highest flood level.

8. *Fixed or Movable / Method of clearance for navigation:* In case of navigable channels which don't have permanent and sufficient clear waterway, fixed or movable bridges are used namely high - level fixed, movable bascule, movable swing or transporter bridge.
9. *Type of loadings:* As per the Indian Road Congress (IRC), road bridges and culverts may be classified into Class A, Class B and Class 70 R (tracked or wheeled) according to the loads they are designed to carry.
10. *Span length:* Under this heading bridges are grouped as culverts if span length is less than 8 m, minor bridges for span lengths between 8 m and 30 m, major bridges for span length more than 30 m and long span bridges for span length above 120 m.
11. *Method of connections :* Depending upon the method of connections of various superstructure parts particular in case of steel construction, steel bridges can be classified into pinned connected, welded or riveted bridges.
12. *Degree of Redundancy :* Depending upon the degree of redundancy bridges can be classified as determinate bridges and indeterminate bridges.
13. *Inter span relation :* Under this category bridges can be grouped as simply supported, continuous or cantilever type bridges.
14. *Type of service to be put to :* In accordance with the type of service anticipated and duration of use, bridges can be classified as permanent, temporary, military bridges like pontoon and bailey, etc.
15. *Number of lanes:* In accordance with the number of lanes for which a bridge structure has been constructed, bridges can be classified as single lane, two lanes or three lane bridges, etc.

Choice of Superstructure Type

The basic function of a bridge superstructure is to permit uninterrupted smooth passage of traffic over it and to transmit the loads and to transmit the load and forces to the substructure safely through the bearings. Although it is difficult to stipulate the aesthetic requirements, it should, however, be ensured that the type of superstructure adopted is simple, pleasing to the eye, and blends with the environment. No hard and fast rules can be laid regarding a the economy in cost. The designer should, however, be able to evolve the most economical type of superstructure based on his judgement and experience given the particular conditions prevailing at the particular site at the particular time. The following factors are to be considered while selecting the type of a bridge superstructure.

i. The nature of river or streams
ii. Nature of foundation / founding strata available
iii. The amount and type of traffic
iv. Whether used for navigation purposes
v. Climatic conditions
vi. Hydraulic data
vii. Type of available construction material
viii. Labour available
ix. The available facilities for erections
x. Maintenance provisional
xi. The availability of funds
xii. Time available for construction
xiii. Strategic consideration
xiv. Economic consideration
xv. Aesthetic consideration.

Characteristics and Design Aspects of Various Types of Superstructures

Basic Principle

The superstructure of any bridge must be designed so as to satisfy the geometric and load carrying requirements to be met. The geometric requirements depend on the number and widths of traffic lanes and footpaths for which the deck is to cater to. It also depends on the overall alignment and various horizontal and vertical clearances required above and below the roadway. Once the geometric considerations are decided, the superstructure has to be designed to meet various structural design requirements. These include considerations of strength, stiffness and stability. This first requires estimation of internal forces and moments and displacements which the externally applied forces will cause in the selected scheme and form of the structure and then deciding the final section sizes, and reinforcements (and prestressing, if the structure is prestressed). The former process constitutes 'analysis' and the latter process constitutes 'design'. The analysis is generally done based on elastic behaviour of the structure. The design is then done either on elastic (i.e. working) strength basis or on load-factored (i.e. ultimate) strength basis, also ensuring the serviceability criteria (like limiting the flexural crack-widths, deflections and vibrations). Some structural analysis work is done using empirical formulae. If the structure is very complicated, model analysis is resorted to.

The selection of superstructure type suitable for particular requirements will depend upon the comparative advantages and disadvantages of each type.

There are seven main kinds of bridges: (1) slab / voided slab bridges, (2) girder bridges, (3) arch bridges, (4) cantilever bridges, (5) suspension bridges, (6) cable-stayed bridges, (7) truss bridges and (8) other bridges

Material Used for Constructing Superstructure

The following materials are generally adopted for constructing permanent bridge decks :

a. Reinforced concrete
b. Prestressed concrete
c. Steel
d. Masonry
e. Composite construction using steel and reinforced concrete, or reinforced and prestressed concrete.

Reinforced Concrete Superstructure

The common types of deck sections in this category are :

a. Solid slab
b. Slab and girder (T-beam)
c. Hollow box girder.

The deck arrangements can be simple or continuous spans, or frame, balanced cantilever, free cantilever cast in situ precast or segmental.

Composite and Steel Superstructure

The various types of deck in this category are :

a. Longitudinal and transverse beams, with concrete slab
b. Longitudinal plate girders, transverse beams and concrete slab

c. Longitudinal box girder with concrete slab

d. Steel truss.

The deck arrangement can be simple or continuous spans, or arch or frame. The slab can be non-composite or can be made composite with the longitudinal beams by shear connectors.

Special Superstructures :

a. Orthotropic deck

b. Cable stayed deck

c. Suspension type deck.

Age masonry superstructure are generally of arch type.

The composite type of decks generally involve combination of two dissimilar structural elements. Precast or prefabricated beams are connected to RC slab by special structural elements called shear connectors, acting together as a unit. The beams may be either :

a. Prefabricated structural steel beams/girders or

b. Precast reinforced concrete, or precast prestressed concrete beams.

Comparison of Reinforced Concrete and Prestressed Concrete Superstructures

- In recent years a majority of bridges have been constructed in prestressed concrete. However, reinforced concrete bridges in certain circumstances still, are more economical.
- Prestressed concrete bridges offer a high degree of freedom from cracks, low maintenance, long life; and they are particularly suitable for dynamic loads and vibrations, and offer great resistance to fatigue arising from repetitions of live loading.
- In RC construction, live load normally approximately doubles the stress in steel, while in prestressed concrete this increase is very little. Position of neutral axis in RC decks about which moments of areas from tension and compression side balance remains fixed, but of that in prestressed concrete shifts with variation in applied load.
- The greatest advantage of prestressed concrete for bridges is reduction in weight slender appearance, possibility of precasting in factory conditions either in segments or even in whole spans, possibility of larger spans and freedom from tension and cracking. For long spans, prestressed concrete decks can

be nearly comparable in weight with steel decks. It is not possible to quote a minimum prestressed concrete span which can be most economical, it depends on local conditions.

For short bridges the advantage of using smaller amount of concrete because of prestressing may be unimportant since there is always a certain basic sot which cannot be changed by the use of prestressing. In such cases the usual reinforced concrete construction may be used with equal advantages.

Bridge Substructure and Foundation

The ' substructure ' and ' foundation ' are generally known to be as the ' better half ' of a bridge structure. This arises from the fact that in a number of cases the substructure and foundation prove to be more expensive than the superstructure especially in case of bridges over major rivers where foundations are very deep and the function of the superstructure carrying the design loads safely depends on the support given by the foundation.

Even though the foundations are not seen by people in general or by those using the bridges but great expertise goes into the design and construction of these in varied soil types, deep waters, marine clays, turbulent flowing rivers, etc. The foundations sometimes reach depths as much as 100 m below the bed. Foundations are very important part of the bridge and the strata on which they are founded has to be unyielding and capable of sustaining the loads. The bridge piers and foundations are required to sustain various loads due to traffic, wind, temperature, differential settlement, seismic forces, etc. The protection of the substructure and foundation is also of utmost importance to ensure the safety of the bridge against the scour of the bed due to water current.

The bridge superstructure cannot be conceived in isolation but is required to be visualised as an integral part of superstructure, structurally as well as architecturally. They are required not only to be engineered to withstand the multitude of loads but also are required to be carefully architecture to blend or harmonize with the surroundings from aesthetic point of view. Thus an integral design approach is required to achieve an architecturally pleasing structure without compromising on the structural integrity and economy.

Abutments

There are two basic types of abutments; open type or spill through type and closed type or non-spill through type.

Open type abutments are more economical than closed type since they allow the back fill to spill out in between their columns thereby reducing the active earth pressure. But due to this very reason, the backfill may settle down, should it be washed through under heavy rains. In such a case of deep curtain wall should be provided extending downward from the abutment cap, for at least 1.5 m, below the spill line.

Piers or Bents

The term 'Pier' pertains to an intermediate substructure unit between the abutment. A bent is a pier consisting of one or more columns, in the substructure, unit with or without cap.

Pile Bents

Type are economical but are suitable for stream crossing where floating debris is minimal. They may be competitive in placed waters and in dry beds, where soil is poor and footings laid at shallower depths might settle. The preferable minimum width of cap is 75 cm. Pile bents are unsightly for bridges 6 m or more in height and unsuitable if scour is significant.

Solid Piers: They are located in streams where floating debris and rolling bounders accompany the last current. They are preferably for long spans, and may be supported on spread footings, pile foundations or caissons, as appropriate. A pier wall should have a low thickness/ height ratio, but should not be less than 30 cm thick. This is the minimum thickness that should be used with two planes of reinforcing steel.

Multicolumn Bents and Multi Columns

They generally support dry land structures. They normally replace pile bents when spread footings are recommended. They may be supported on either spread footings or pile foundations. Columns may be circular rectangular or variable in section to provide architectural effect. The thickness/height ratio of normal height uniform section columns should be in a range of 1/12 to 1/15 for appearance as well as structural reasons. Unnecessarily thick columns can invite high temperature and shrinkage stresses and very slender ones will require buckling analysis.

Single Column Bents and Single Columns

They can sometimes be used to avoid skewed piers. They are adapted to viaducts over city streets where the location of the column may be restricted.

Foundations

Footings are appropriate where scour is low and soil bearing capacity is enough not to create settlement problems, or if rock is available at a shallow depth. In any case, if subsoil water is encountered footings are feasible only if dewatering is possible. Otherwise piles are preferable, which enables transferring the loads directly to lower and firmer substrata, without requiring dewatering. These are costlier than footings, in general. However, in case of deep scour, rolling bounders, floating debris and strong current, the (generally costlier) caissons become necessary. Caissons are relatively massive and generally suit spans not less than 25 m preferably longer.

Choice of Type of Foundation

The choice of foundation for the bridges depends upon the conditions of soil. Cost of substructure increases with the depth of foundation. Similarly, the cost of superstructure per unit length also increases with the span length. Hence the decision on span lengths to depend upon the ratio of the cost of substructure including the foundation versus the cost of superstructure i.e. the girder and the deck and it can be found that it is most economical when the ratio is one. There may be cases where though the cost of foundations may be more than the cost of superstructure practical difficulties in sinking the foundation and limitations may require a reconsideration for the reduction of the total number of foundations. Navigational considerations also will have an influence upon the span. The choice of span has to be suitably made taking all these into consideration.

Foundations can be broadly classified as :

(i) open foundations;

(ii) pile foundations; and

(iii) well foundations.

Open foundations are suitable for small bridges and bridges which re sited in a location where rock or firm founding soil is available at a shallow depth and, at the same time, the scour in the river is not considerable. The next choice, from considerations of economy, are the pile foundation which are again more suitable for being used in cohesive soils and beds subject to minimum scour. Well foundations have been more popularly used in India in view of the expertise available particularly for major and important bridges.

General Design Procedure for Design of Substructure of A Bridge

I. First of all compute the dead load reaction due to the superstructure and superimposed dead load along with the maximum and minimum live load reactions coming on the pier or abutment under consideration. Apart from this also compute the corresponding moments caused due to these loads about the longitudinal as well as the transverse directions or axis of the bridge. For this the critical case is the maximum possible eccentric transfer of these dead and live load reactions along with the braking and temperature forces, etc. to the foundation base.

II. Compute the total vertical reaction at the base of the foundation or at the soft of the pile cap in case of pile foundation under maximum, minimum and no live load conditions, taking into account the upward force due to the effect of buoyancy equal to 50% or 100% weight of water in volume equal to that of the submerged mass. This value is 100% in case of saturated soil and fissured, weathered or weak rock and 50% in case of good rock.

III. Compute the total moment about the bridge's longitudinal axis and the horizontal force in the transverse direction, at various levels and more importantly at the base of the foundation due to possible eccentricity of live and dead loads, force due to water current and afflux, if any.

IV. Compute the total moment about the bridge's transverse axis and the horizontal force in the longitudinal direction, at various levels and more importantly at the base of the foundation due to possible eccentricity of live and dead loads, force due to braking and temperature, water current and afflux, if any. The force due to braking can be calculated depending upon the following cases : (i) In case of simply supported spans on rocker / roller bearings, braking force due to live load on one complete span may be assumed to go to its rocker bearing only such that the corresponding foundation under it will resist force either equal to braking - temperature or half of braking + temperature, the latter case being critical for usually an abutment or a pier supporting unequal spans. (ii) In case of simply supported spans with identical neoprene or elastomeric bearings under each of an individual span, the foundation will resist the sum of half of the value of braking force due to live loads on each of the two spans supported by it and the horizontal temperature force equal to the difference of the product of the shear rating of the neoprene bearings and the deck movements above them for left and right hand side spans respectively. (iii) In case of continuous bridge structures the horizontal force being resisted by the bearings depends upon the arrangement and type of bearings chosen and is case specific.

V. Compute the wind force in the transverse direction acting on the exposed area of the bridge deck and substructure above the lowest bed level. This force is calculated with and without the live loading on the bridge deck. Usually it is sufficient to consider wind on the deck surface area between its soffit and top of the solid parapet or crash barrier, and that of the body of the live load at the given rate as per codal recommendations. Similarly for no live load condition the rate specified by the code is considered. In certain cases the wind force strikes the bridge structure obliquely. In such a case a combination of simultaneously acting wind forces acting in the transverse and longitudinal directions is considered as per codal specifications.

VI. Compute (i) the static equivalent of the horizontal seismic force depending upon the mass of the structure above the earth embedment level which is the maximum scour level in case of hydraulic bridges. The seismic forces are based on the full weight including the submerged portions as long as they are above the embedment level. (ii) The seismic force on a mass can be taken as a certain fraction of its weight, acting through its centroid. This fraction varies depending upon the bridge site being in a severe seismic zone to a non - seismic area as per recommendations of the code. Apart from this a vertical seismic force acting in the upward or downward direction and equal to half of the above mentioned force. This vertical force becomes critical in cases where values of maximum and minimum base pressures are critical. (iii) In case of seismic force acting in the transverse direction, the live load seismic effect is assumed to be acting at 1.2 m above the formation level and that due to the dead load of the deck and superimposed dead load at the deck centroid level. (iv) In case the seismic force is in the direction of the longitudinal axis of the bridge, then the seismic effect due to live load is to be ignored since the braking force is already being considered and any further longitudinal force on the live load will only cause skidding of the wheels.

VII. Estimate the 'active' earth pressure force and moment (at various levels and at the base of foundation) on account of the earthfill retained upto foundation base in case of founding strata being soil or upto rock level in case of rocky founding strata. Passive relief from the front fill is generally to be ignored, but if it is well protected then it may be considered depending upto the actual conditions in situ (but accounting for the negative surcharge angle effect if sloping downwards). In case of wells where the foundation is usually taken down well into permanent and unexecuted soil (e.g. a caisson taken below the maximum scour level), estimate and take into account the net 'passive less active' earth pressure relief (force and moment) as a result of the soil grip.

VIII. Compute the net vertical load, the net horizontal forces in the longitudinal and transverse directions and the net moments about these two directions, at the base of the foundation (at soffit of pile cap in the case of piles) under each critical load combination. Apart from the pile foundations, calculate the base pressures and the safety factors for stability against overturning and sliding and ensure that the requirements are satisfied. In case one of the parametres is not satisfied, then redesign by taking a new set of dimensions.

IX. In case of piled foundations estimate the maximum and minimum axial loads in the piles by the using rivet-group approach. Ensure that no pile is in tension unless tension-piles are provided. Ensure the structural design of the pile-section for the vertical loads and moments combinations on it. Check for the adequacy of soil resistance around an individual pile, the block failure and group action of the pile group.

X. After satisfying the stability requirements of the foundation, calculate the vertical load and moments at various intermediate critical sections (including at the maximum bending moment section in a caisson) in the foundation structure (this includes whole pier/abutment structure), and then structurally design these sections. Also design various other structural components of the foundations structure, taking into account the respective critical load combinations for these components.

XI. Estimate the vertical loads on each bearing under various load combinations, along with the coexisting longitudinal and transverse horizontal forces and rotations. Design the bearings for these effects. Use may be made of the standard manufacturers' catalogues for certain standard bearings.

Types of Piers and their Design Aspects

Open Piers : They are classified as :

1. Multiple bent or multiple column
2. Pile bent

3. Cylindrical piers
4. Trestle piers.

Multiple Bent : They are often used on ground. The section of wall at the bottom may or may not be used. It is needed in overpass work where traffic runs parallel close to the bent to reduce damage to the columns in case of accident. It is lighter and may be more economical than the solid pier; but it requires more formwork and material.

Pile Bent : The pile bent is used for low piers over unstable ground. They consist of reinforced concrete or steel piles driven into the ground, provided with a capping at their top to support the main girders. They are laterally connected by reinforced concrete or steel braces. The pile is used both for a support by driving to resistance and for a column by projecting above ground. The outside piles are usually battered.

Cylindrical Piers : They consist of mild steel or cast iron cylinders which are filled with concrete. When cylinders are used apart and suitable bracing is provided.

Trestle Piers : They are used for temporary work and for timber work. They are made up of RCC or steel vertical, horizontal and diagonal members. The connections in steel trestle may be either riveted or welded. RCC trestle type pier consists of circular columns with a bent cap at the top. In order to avoid moments being transferred from deck to the columns concrete hinges are introduced between the top of the columns and the bent cap. In case of fly overs and elevated roads where trestle piers are used, it is desirable to provide connecting diaphragms between the column.

The following are some of the special piers :

1. Separate piers
2. Abutment piers
3. Cellular piers
4. Framed piers.

Separate piers : They are sometimes used for high bridges. These may be low piers with steel columns above on concrete piers upto the bridge level. In water, they may be combined with a solid shaft below upto flood water level. They may have the advantage of lower cost than solid shafts.

Abutment Piers : In the case of multiple span arch bridges, every third or fourth pier is designed as an abutment to receive the thrust from either side. Such a pier is thicker in section and is known as abutment pier. The object of the abutment pier is listed below :

i. The arches can be turned in sets, and thus heavy expenditure in centreing is saved

ii. In the event of damage, occurring to any portion of the bridge due to floods, the same will probably not extend further than the next abutment.

Cellular Piers: The cellular piers consists of two concentric RCC Shells connected by radial ribs and horizontal bands at suitable intervals. The intermediate space between the shell is filled with some filler material. This will result in saving of cement concrete which is a costly material. This type of piers usually require difficult shuttering and additional labour in placing reinforcements.

Framed Piers : This type of piers result in reduced effective span length for girders on ether side of the centre line of the pier leading to economy in the cost of superstructure. In this type of design two expansion joints at close intervals to 1 to 2 m are provided which results in uneven surface and maintenance problems. Such type of piers if used in rivers subjected to sudden floods near hills and forests would be subjected to floating debris e.g. floating tree trunk etc.

Design of Piers

The following loads and forces are taken into account :

- Dead weight of the Pier
- End reactions due to dead and live loads of superstructure : These reactions will cause an eccentricity when one span only is loaded. Also some end details are so standardised as to dimension that eccentricity which may be caused by difference in span lengths.
- Impact forces due to floating debris and live load for the top 3 m of the pier only.
- Wind pressure acting on the superstructure, moving loads and the pier.
- Centrifugal forces in the event of a road bridge situated on a curve.
- Longitudinal forces arising either due to tractive effort caused through acceleration of the vehicle or braking effect are resulting from the application of the brakes and due to friction at bearings.
- Water pressure

- Earthquake forces in the event of bridge located in seismic zone.
- Buoyancy effect of submerged part of substructure.

The magnitude of the above forces can be determined as per IRC Standard specifications and code of practice section II. Bridge spanning across a tidal river or an estuary, are subjected to an additional hydrodynamic forces due to wave action. The wave motion is essentially an unsteady fluid flow. The fluid particles are subjected to not only a velocity in the horizontal and vertical directions but also to acceleration in these directions. Thus, the fluid particles velocity and acceleration induce a horizontal drag force and an inertial force. The forces developed in the vertical direction are insignificant because of the piers are normally rigidly fixed at the bottom.

Stability of Bridge Piers

Bridge piers are considered stable if they satisfy the following conditions :

1. All the components parts of the pier must be sufficiently strong to resist bending moments and shearing forces.
2. The pier should be stable against sliding downstream and sliding parallel to the axis of the bridge.
3. The pier must not be turned over about the downstream toe and also in the direction of the axis of the bridge.
4. Under the worst combinations of all loads and forces, no tension should be permitted under the foundations, and the maximum pressure should not exceed the safe bearing capacity of the soil.

3

Construction Surveying

Construction surveying (otherwise known as "layout" or "setting-out") is to stake out reference points and markers that will guide the construction of new structures such as roads or buildings. These markers are usually staked out according to a suitable coordinate system selected for the project.

History of Construction Surveying

- The nearly perfect squareness and north-south orientation of the Great Pyramid of Giza, built c. 2700 BC, affirm the Egyptians' command of surveying.
- A recent reassessment of Stonehenge (c. 2500 BC) suggests that the monument was set out by prehistoric surveyors using peg and rope geometry.
- In the sixth century BC geometric based techniques were used to construct the tunnel of Eupalinos on the island of Samos.

Elements of the Construction Survey

- Survey existing conditions of the future work site, including topography, existing buildings and infrastructure, and underground infrastructure whenever possible (for example, measuring invert elevations and diametres of sewers at manholes);
- Stake out reference points and markers that will guide the construction of new structures;
- Verify the location of structures during construction;
- Conduct an As-Built survey: a survey conducted at the end of the construction project to verify that the work authorised was completed to the specifications set on plans.

Coordinate Systems used in Construction

Land surveys and surveys of existing conditions are generally performed according to geodesic coordinates. However for the purposes of construction a more suitable coordinate system will often be used. During construction surveying, the surveyor will often have to convert from geodesic coordinates to the coordinate system used for that project.

Chainage or Station

In the case of roads or other linear infrastructure, a *chainage* will be established, often to correspond with the centre line of the road. During construction, structures would then be located in terms of *chainage*, *offset* and *elevation*. *Offset* is said to be "left" or "right" relative to someone standing on the *chainage line* who is looking in the direction of increasing *chainage*. Plans would often show *plan* views (viewed from above), *profile* views (a "transparent" section view collapsing all section views of the road parallel to the *chainage*) or *cross-section* views (a "true" section view perpendicular to the *chainage*). In a *plan* view, *chainage* generally increases from left to right, or from the bottom to the top of the plan. *Profiles* are shown with the chainage increasing from left to right, and *cross-sections* are shown as if the viewer is looking in the direction of increasing *chainage* (so that the "left" *offset* is to the *left* and the "right" *offset* is to the *right*).

"Chainage" may also be referred to as "Station".

Building Grids

In the case of buildings, an arbitrary system of grids is often established so as to correspond to the rows of columns and the major load-bearing walls of the building. The grids may be identified alphabetically in one direction, and numerically in the other direction (as in a road map). The grids are usually but not necessarily perpendicular, and are often but not necessarily evenly spaced. Floors and basement levels are also numbered. Structures, equipment or architectural details may be located in reference to the floor and the nearest intersection of the arbitrary axes.

Other Coordinate Systems

In other types of construction projects, arbitrary "north-south" and "east-west" reference lines may be established, that do not necessarily correspond to true coordinates.

Equipment and Techniques used in Construction Surveying

Surveying equipment, such as levels and theodolites, are used for accurate measurement of angular deviation, horizontal, vertical and slope distances. With computerisation, electronic distance measurement (EDM), total stations, GPS surveying and laser scanning have supplemented (and to a large extent supplanted) the traditional optical instruments. The builder's level measures neither horizontal nor vertical angles. It simply combines a spirit level and telescope to allow the user to visually establish a line of sight along a level plane. When used together with a graduated staff it can be used to transfer elevations from one location to another.

An alternative method to transfer elevation is to use water in a transparent hose as the level of the water in the hose at opposite ends will be at the same elevation.

Equipment and Techniques used in Mining and Tunneling

Total stations are the primary survey instrument used in mining surveying.

Underground Mining

A total station is used to record the absolute location of the tunnel walls (stopes), ceilings (backs), and floors as the drifts of an underground mine are driven. The recorded data is then downloaded into a CAD program, and compared to the designed layout of the tunnel.

The survey party installs control stations at regular intervals. These are small steel plugs installed in pairs in holes drilled into walls or the back. For wall stations, two plugs are installed in opposite walls, forming a line perpendicular to the drift. For back stations, two plugs are installed in the back, forming a line parallel to the drift.

A set of plugs can be used to locate the total station set up in a drift or tunnel by processing measurements to the plugs by intersection and resection.

Professional Status of Construction Surveyors

Building Surveying emerged in the 1970's as a profession in the United Kingdom by a group of technically minded General Practice Surveyors. Building Surveying is a recognised profession within Britain and Australia.

In Australia in particular, due to risk mitigation/limitation factors the employment of surveyors at all levels of the construction industry is widespread. There are still many countries where it is not widely recognised as a profession. The Services that Building Surveyors undertake are broad but include:

- Construction design and building works
- Project Management and monitoring
- CDM Co-ordinator under the Construction (Design & Management) Regulations 2007
- Property Legislation adviser
- Insurance assessment and claims assistance
- Defect investigation and maintenance adviser
- Building Surveys and measured surveys
- Handling Planning applications
- Building Inspection to ensure compliance with building regulations
- Undertaking pre-acquisition surveys
- Negotiating dilapidations claims.

Building Surveyors also advise on many aspects of construction including:

- design
- maintenance
- repair
- refurbishment
- restoration.

Clients of a building surveyor can be the public sector, Local Authorities, Government Departments as well as private sector organisations and work closely with architects, planners, homeowners and tenants groups. Building Surveyors may also be called to act as an expert witness. It is usual for building surveyors to undertake an accredited degree qualification before undertaking structured training to become a member of a professional organisation. For Chartered Building Surveyors, these courses are accredited by the Royal Institution of Chartered Surveyors. Other professional organisations that have building surveyor members include CIOB, ABE, HKIS and RICS.

With the enlargement of the European community, the profession of the Chartered Building Surveyor is becoming more widely known in other European states, particularly France. Chartered Building Surveyors, where many English speaking people buy second homes.

Distinction from Land Surveyors

In the United States, Canada, the United Kingdom and most Commonwealth countries land surveying is considered to be a distinct profession. Land surveyors have their own professional associations and licencing requirements. The services of a licenced land surveyor are generally required for boundary surveys (to establish the boundaries of a parcel using its legal description) and subdivision plans (a plot or map based on a survey of a parcel of land, with boundary lines drawn inside the larger parcel to indicated the creation of new boundary lines and roads).

Transport Engineering

Transportation engineering is the application of technology and scientific principles to the planning, functional design, operation and management of facilities for any mode of transportation in order to provide for the safe, rapid, comfortable, convenient, economical, and environmentally compatible movement of people and goods (transport). It is a sub-discipline of civil engineering. Transportation engineering is a major component of the civil engineering discipline. The importance of transportation engineering within the civil engineering profession can be judged by the number of divisions in ASCE (American Society of Civil Engineers) that are directly related to transportation. There are six such divisions (Aerospace; Air Transportation; Highway; Pipeline; Waterway, Port, Coastal and Ocean; and Urban Transportation) representing one-third of the total 18 technical divisions within the ASCE (1987).

Figure : *The engineering of this roundabout in Bristol, England, attempts to make traffic flow free-moving.*

The planning aspects of transport engineering relate to urban planning, and involve technical forecasting decisions and political factors. Technical forecasting of passenger travel usually involves an urban transportation planning model, requiring the estimation of trip generation (how many trips for what purpose), trip distribution (destination choice, where is the traveller going), mode choice (what mode is being taken), and route assignment (which streets or routes are being used). More sophisticated forecasting can include other aspects of traveller decisions, including auto ownership, trip chaining (the decision to link individual trips together in a tour) and the choice of residential or business location (known as land use forecasting). Passenger trips are the focus of transport engineering because they often represent the peak of demand on any transportation system.

A review of descriptions of the scope of various committees indicates that while facility planning and design continue to be the core of the transportation engineering field, such areas as operations planning, logistics, network analysis, financing, and policy analysis are also important to civil engineers, particularly to those working in highway and urban transportation. The National Council of Examiners for Engineering and Surveying (NCEES) list online the safety protocols, geometric design requirements, and signal timing.

Transportation engineering, as practised by civil engineers, primarily involves planning, design, construction, maintenance, and operation of transportation facilities. The facilities support air, highway, railroad, pipeline, water, and even space transportation. The design aspects of transport engineering include the sizing of transportation facilities (how many lanes or how much capacity the facility has), determining the materials and thickness used in pavement designing the geometry (vertical and horizontal alignment) of the roadway (or track). Before any planning occurs the Engineer must take what is known as an inventory of the area or if it is appropriate, the previous system in place. This inventory or database must include information on (1) population, (2)l and use, (3) economic activity, (4) transportation facilities and services, (5) travel patterns and volumes, (6) laws and ordinances, (7) regional financial resources, (8) community values and expectations. These inventories help the engineer create business modells to complete accurate forecasts of the future conditions of the system Review.

Operations and management involve traffic engineering, so that vehicles move smoothly on the road or track. Older techniques include signs, signals, markings, and tolling. Newer technologies involve intelligent transportation systems, including advanced traveller information systems (such as variable message signs), advanced traffic control systems (such as ramp metres), and vehicle infrastructure integration. Human factors are an aspect of transport engineering, particularly concerning driver-vehicle interface and user interface of road signs, signals, and markings.

Engineers in this specialisation:

- Handle the planning, design, construction, and operation of highways, roads, and other vehicular facilities as well as their related bicycle and pedestrian realms.
- Estimate the transportation needs of the public and then secure the funding for the project.
- Analyse locations of high traffic volumes and high collisions for safety and capacity.
- Use civil engineering principles to improve the transportation system.
- Utilises the three design controls which are the drivers, the vehicles, and the roadways themselves.

Railroad Engineering

Railway engineers handle the design, construction, and operation of railroads and mass transit systems that use a fixed guideway (such as light rail or even monorails). Typical tasks would include determining horizontal and vertical alignment design, station location and design, and construction cost estimating. Railroad engineers can also move into the specialised field of train dispatching which focuses on train movement control.

Railway engineers also work to build a cleaner and safer transportation network by reinvesting and revitalising the rail system to meet future demands. In the United States, railway engineers work with elected officials in Washington, D.C. on rail transportation issues to make sure that the rail system meets the country's transportation needs.

Port and Harbour Engineering

Port and harbour engineers handle the design, construction, and operation of ports, harbours, canals, and other maritime facilities. This is not to be confused with marine engineering.

Airport Engineering

Airport engineers design and construct airports. Airport engineers must account for the impacts and demands of aircraft in their design of airport facilities. These engineers must use the analysis of predominant wind direction to determine runway orientation, determine the size of runway border and safety areas, different wing tip to wing tip clearances for all gates and must designate the clear zones in the entire port.

Municipal or Urban Engineering

Municipal or urban engineering applies the tools of science, art and engineering in an urban environment. Municipal engineering is concerned with municipal infrastructure. This involves specifying, designing, constructing, and maintaining streets, sidewalks, water supply networks, sewers, street lighting, municipal solid waste management and disposal, storage depots for various bulk materials used for maintenance and public works (salt, sand, etc.), public parks and bicycle paths. In the case of underground utility networks, it may also include the civil portion (conduits and access chambers) of the local distribution networks of electrical and telecommunications services. It can also include the optimising of garbage collection and bus service networks. Some of these disciplines overlap with other civil engineering specialities, however municipal engineering focuses on the coordination of these infrastructure networks and services, as they are often built simultaneously (for a given street or development project), and managed by the same municipal authority.

History

Modern municipal engineering finds its origins in the 19th century United Kingdom, following the Industrial Revolution and the growth of large industrial cities. The threat to urban populations from epidemics of waterborne diseases such as cholera and typhus lead to the development of a profession devoted to "sanitary science" that later became "municipal engineering".

A key figure of the so-called "public health movement" was Edwin Chadwick, author of the parliamentary report "The Sanitary Condition of the Labouring Population", published in 1842.

Early British legislation included:

- Burgh Police Act 1833 - powers of paving, lighting, cleansing, watching, supplying with water and improving their communities.
- Municipal Corporations Act 1835
- Public Health Act 1866 – formation of drainage boards

- Public Health Act 1875 known at the time as the Great Public Health Act.

This legislation provided local authorities with powers to undertake municipal engineering projects and to appoint borough surveyors (later known as "municipal engineers").

In the U.K, the Association of Municipal Engineers, (subsequently named Institution of Municipal Engineers), was established in 1874 under the encouragement of the Institution of Civil Engineers, to address the issue of the application of sanitary science. By the early 20th century, Municipal Engineering had become a broad discipline embracing many of the responsibilities undertaken by local authorities, including roads, drainage, flood control, coastal engineering, public health, waste management, street cleaning, water supply, sewers, waste water treatment, crematoria, public baths, slum clearance, town planning, public housing, energy supply, parks, leisure facilities, libraries, town halls and other municipal buildings.

In the UK, the development of different strands of knowledge necessary for the management of municipal infrastructure led to the emergence of separate specialised institutions, including:

- For drainage: Chartered Institution of Water and Environmental Management, 1895
- For town planning: Town Planning Institute 1914 ... subsequently becoming the Royal Town Planning Institute
- For street lighting: Association of Public Lighting Engineers, 1934 subsequently becoming the Institution of Lighting Engineers
- For highway engineering: Institution of Highways and Transportation, 1930
- For public housing: Institute of Housing, 1931.

In 1984 the Institution of Municipal Engineers merged with the Institution of Civil Engineers. Since the 1970's, there has been a global trend toward increasing privatisation and outsourcing of municipal engineering services. In the UK in the 1990's a change in management philosophy brought the demise of the traditional organisational structure of boroughs where the three functions of town clerk, borough treasurer and borough engineer were replaced by an administrative structure with a larger number of specialised departments. In the late 1990's and early 21st century there was increasing dissatisfaction over what was perceived to be fractured and dysfunctional public services designed along narrow specialities. A more holistic approach to urban engineering began to emerge as an alternative concept. Critics of the specialised approach included the Commission for Architecture and the Built Environment that complained that the specialised approach to management of the public realm focused too much on the on efficient movement of vehicles rather than the more general interests of local communities.

Professional Practice

In the United Kingdom there is no longer any formal professional qualification in municipal engineering although there are degree courses available in urban engineering. A professional certificate in Urban Engineering is available from through the Institution of Incorporated Engineers via the Public Realm Information and Advice Network.

The British Institution of Civil Engineers (ICE) caters to practitioners employed in the public sector, private consultancy and academia through its Proceedings Journal Municipal Engineer. The journal, first published in 1873, has a global scope and covers the whole life cycle of municipal services addressing technical, political and community issues. In addition an Expert Panel responds on behalf of ICE to Government consultations and is represented on the International Federation of Municipal Engineering.

International Organisation

The International Federation of Municipal Engineering (IFME) is an organisation comprising professional municipal engineers from all round the world. IFME's mission is to connect municipal engineers, public works professionals, public agencies, institutions and businesses around the world in order that they can share a global pool of knowledge and experience. The aim is to foster continued improvement in the quality of public works and wider community services.

The inaugural meeting was held in 1960 at the UNESCO headquarters in Paris. Membership has grown steadily and in 2009 comprised representatives from national associations in: Australia, Canada, Denmark, Estonia, Finland, Italy, Israel, The Netherlands, New Zealand, Norway Southern Africa, Sweden and the UK. Belgium and San Marino are presently Corresponding Members.

Related Engineering Disciplines

Municipal or urban engineering combines elements of environmental engineering, water resources engineering and transport engineering.

Relationship to Urban Design or Urban Planning

Today, municipal engineering may be confused with urban design or urban planning. Whereas the urbanist or urban planner may design the general layout of streets and public places, the municipal engineer is concerned with the detailed design. For example, in the case of the design of a new street, the urbanist may specify the general layout of the street, including landscaping, surface finishings and urban accessories, but the municipal engineer will prepare the detailed plans and specifications for the roads, sidewalks, municipal services and street lighting. However Municipal Engineering as practised a century ago fully embraced the function of urban design and urban planning, even though the terms had yet to be coined.

Traffic Engineering (Transportation)

Traffic engineering is a branch of civil engineering that uses engineering techniques to achieve the safe and efficient movement of people and goods on roadways. It focuses mainly on research and construction of the infrastructure necessary for safe and efficient traffic flow, such as road geometry, sidewalks and crosswalks, segregated cycle facilities, shared lane marking, traffic signs, road surface markings and traffic lights.

Traffic engineering is closely associated with other disciplines:

- Transport engineering
- Pavement engineering
- Bicycle transportation engineering
- Highway engineering
- Transportation planning
- Urban planning
- Human factors engineering.

Typical traffic engineering projects involve designing traffic control device installations and modifications, including traffic signals, signs, and pavement markings. However, traffic engineers also consider traffic safety by investigating locations with high crash rates and developing countermeasures to reduce crashes. Traffic flow management can be short-term (preparing construction traffic control plans, including detour plans for pedestrian and vehicular traffic) or long-term (estimating the impacts of proposed commercial developments on traffic patterns). Increasingly, traffic problems are being addressed by developing systems for intelligent transportation systems, often in conjunction with other engineering disciplines, such as computer engineering and electrical engineering.

Traffic Systems

Traditionally, road improvements have consisted mainly of building additional infrastructure. However, dynamic elements are now being introduced into road traffic management. Dynamic elements have long been used in rail transport. These include sensors to measure traffic flows and automatic, interconnected, guidance systems to manage traffic (for example, traffic signs which open a lane in different directions depending on the time of day). Also, traffic flow and speed sensors are used to detect problems and alert operators, so that the cause of the congestion can be determined, and measures can be taken to minimize delays. These systems are collectively called intelligent transportation systems.

Lane Flow Equation

The relationship between lane flow (Q, vehicles per hour), maximum speed (V, kilometres per hour) and density (K, vehicles per kilometre) is:

$$Q = KV$$

Observation on limited access facilities suggests that up to a maximum flow, speed does not decline while density increases. However, above a critical threshold, increased density reduces speed. Additionally, beyond a further threshold, increased density reduces flow as well. Therefore, speeds and lane flows at bottlenecks can be kept high during peak periods by managing traffic density using devices that limit the rate at which vehicles can enter the highway. Ramp metres, signals on entrance ramps that control the rate at which vehicles are allowed to enter the mainline facility, provide this function (at the expense of increased delay for those waiting at the ramps).

Highway Safety

Highway safety engineering is a branch of traffic engineering that deals with reducing the frequency and severity of crashes. It uses physics and vehicle dynamics, as well as road user psychology and human factors engineering, to reduce the influence of factors that contribute to crashes.

A typical traffic safety investigation follows these steps

1. Identify and priorities investigation locations. Locations are selected by looking for sites with higher than average crash rates, and to address citizen complaints.

2. Gather data. This includes obtaining police reports of crashes, observing road user behaviour, and collecting information on traffic signs, road surface markings, traffic lights and road geometry.
3. Analyse data. Look for collisions patterns or road conditions that may be contributing to the problem.
4. Identify possible countermeasures to reduce the severity or frequency of crashes.
   - Evaluate cost/benefit ratios of the alternatives
   - Consider whether a proposed improvement will solve the problem, or cause "crash migration." For example, preventing left turns at one intersection may eliminate left turn crashes at that location, only to increase them a block away.
   - Are any disadvantages of proposed improvements likely to be than the problem you are trying to solve?
5. Implement improvements.
6. Evaluate results. Usually, this occurs some time after the implementation. Have the severity and frequency of crashes been reduced to an acceptable level? If not, return to step 2.

Bicycle Transportation Engineering

Bicycle transportation engineering is the study of transportation engineering as it affects bicycles and cycling. It includes the design of dedicated transport facilities for cyclists, but also the study of how mixed-mode environments (i.e. where cyclists are not segregated from other traffic) can be made to work safely.

Roads

Various methods of altering or reallocating the roadway right-of-way to facilitate bicycling and create bikeways have been added to many of the manuals used by transport planners and engineers.

Segregated Cycle Facilities

Sidepath or Shared use Path: A shared use path is a bikeway that is physically separated from motorised vehicle traffic by an open space (generally landscaping) or some form of barrier.

Cycle Track

A cycle track, also called a *separated cycle lane,* is a bicycle exclusive facility that provide physical separation from motorised vehicle traffic within the road right-of-way, generally by means of kerbs or other barriers. Cycle tracks can either incorporate bicycle-only signal phases at intersections (for 100% separation) or utilise "mixing zones" to merge bicycle and motor vehicle traffic. A cycle track combines the user experience of a separated path with the on-road infrastructure of a bike lane.

Bicycle Lane

A designated bicycle lane, according to the 1998 United States Manual on Uniform Traffic Control Devices, is:

- a portion of a roadway which has been designated for use by bicyclists
- marked by a white (usually solid) stripe painted on the pavement
- significantly narrower than traffic lanes
- found at the side of the traffic lanes.

Cycle lanes require less space compared with the option of providing side paths or separate tracks, and generally do not have the legal / priority issues that facilities separated from the general traffic may have (depending on the locally applicable road laws). However, if traffic is heavy or fast, the lack of physical separation from motor vehicles may make cycle lanes unattractive to less confident cyclists. The lack of separation also means they are more apt to be blocked by (illegally) parked vehicles.

Shoulder

A hard shoulder, or simply shoulder, is a reserved area outside of a roadway, but within the road right-of-way. In addition to (a) serving as a buffer area in the event of a temporary loss of vehicle control, (b) providing emergency access for ambulances and police cars and (c) providing a space for inoperable vehicles, it may also be used by pedestrians and bicyclists.

Non-segregated Facilities

Wide Outside Lane (WOL): Wide outside through lanes help motorists pass slower cyclists without having to decrease speed or change lanes. This method is held by some to be particularly important on roads with a high proportion of wide vehicles, such as buses or heavy trucks. These lanes also provide more room for cyclists to filter past queues of other vehicles in congested conditions. The use of such lanes is specifically endorsed by *Cycling: the way ahead for towns and cities*, the European Commission policy document on cycle promotion.

Cycle friendly infrastructure argues for a marked lane width of 4.25 m (14 ft). It is argued that, on undivided roads, this width provides cyclists with adequate clearance from passing wide vehicles while being sufficiently narrow to deter motorists from attempting to "double up" and form two lanes. This "doubling up" effect may be related to junctions. At non-junction locations, greater width might be preferable if this effect can be avoided.

Shared Bus Lane

A shared bus lane is a bus lane that allows cyclists to use it. Depending on the width of the lane, the speeds and number of buses, and other local factors, the safety and popularity of this arrangement vary. Guidance produced for Cycling England endorses bus lanes as providing cyclists with a *direct and barrier free route into town centres* and as avoiding the difficulties associated with other provisions such as shared-use footways. According to a French survey 42% of cyclists described themselves as "enthusiasts" for shared bus bike lanes versus 33% who were of mixed opinion and 27% who were opposed. Many cycling activists view these as being more attractive than cycle paths, while others object to being in close proximity to bus exhausts. The sharing of bus lanes has been described as "generally very popular" with cyclists in London.

As of 2003, mixed bus/cycle lanes accounted for 118 km of the 260 km of cycling facilities in Paris. The French city of Bordeaux has 40 km of shared bus cycle lanes. It is reported that that in the city of Bristol, a showcase bus priority corridor, where road space was reallocated along a 14 km stretch also resulted in more space for cyclists and had the effect of increasing cycling. The reverse effect has also been suggested, a review carried out in London reports that cycling levels fell across Kew bridge following the removal of a bus lane - this was despite a general increase in cycling level in the city generally. In addition, it is arguably easier, politically speaking, to argue for funding of joint facilities rather than the additional expense of both segregated cycling facilities and bus-only lanes.

In some instances bus lane proposals have run into vehement opposition from cycling representatives - a typical theme is the perceived generation of conflict due to the narrowing of other lanes already shared by cars/cyclists so as to create space for the bus lanes The TRL reports that cyclists and bus drivers tend to have low opinions of each other There have been reports in Dublin of conflict as cyclists choose to cycle in the bus lanes and a bus driver apparently expected them to use adjacent cycle tracks instead. In some other cities the arrangements seem to work successfully, with bus companies and cyclists' groups taking active steps to ensure that understanding is improved between the two groups of road users.

Streets

Shared Space: Shared space schemes, which are characterised by the removal of road markings, signs and signals, give all street users equal priority and equal responsibility for each others safety. Experiences where these schemes are in use, show that street users, particularly motorists, undirected by signs, curbs or road markings, reduce their speed and establish eye contact with other users. Results from the thousands of such implementations worldwide all show casualty reductions and most also show reduced journey times. Following the partial conversion of London's Kensington High Street to *shared space*, accidents were reduced there by 44% (the London average was 17%).

Bicycle Boulevards

A bicycle boulevard is a low speed street which has been designed to discourage cut-through motor vehicle traffic and to give priority to cyclists as through-going traffic.

Bike Paths

Bike paths are bicycle routes that follow an independent right-of-way. There are two distinct types of bike paths: those used exclusively by bicycles and those shared with pedestrians. In the United States almost all bike paths are shared with pedestrians. Bike paths are often built on old railroad right-of-ways (called rail trails) and as part of greenways and fore shoreways.

Traffic Lights

How traffic signals are designed and implemented directly impacts cyclists. For instance, poorly adjusted vehicle detector systems, used to trigger signal changes, may not correctly detect cyclists. Traffic managers in Copenhagen link cyclist-specific traffic signals on a major arterial bike lane to provide green waves for rush hour cycle-traffic. Cycling-specific measures that can be applied at traffic signals include the use of advanced stop lines and/or bypasses. The frequency with which lights change is important to cyclists who may conserve energy by anticipating green lights ahead i.e. the shorter the interval the better for cyclists. Intersection clearance times and green wave timing may also be relevant.

Road Surface

Bicycle tires being narrow, road surface is more important than for other transport, for both comfort and safety. The type and placement of storm drains, manholes, surface markings, and the general road surface quality should all be taken into account by a bicycle transportation engineer. Drain grates, for example, must not catch wheels.

Parking

Bicycle parking is another important part of Bicycle Transportation Engineering. In most of the United States, bicycle parking facilities are scarce, or are so inadequate that nearby trees or parking metres are used. The hitching post type of bicycle stand is an improvement over the old type that had a slot for the front wheel but only allow for two bicycles per post. The Netherlands, where bicycles are much in use, has two-tiered bicycle racks giving high density (the handlebars overlap) and security (the bicycle is held well and is easy to lock).

Secure bicycle parking is argued to be a key factor influencing the decision to cycle. To be considered secure, the parking must be of a suitable design: allowing the bicycle to be locked via the frame. A readily observable location can also permit so-called passive security from passers-by. Weather protection is also desirable. As a rule, where cycling is encouraged as an alternative to motoring, efforts are made to make bicycle parking more convenient and attractive to use than nearby car parking arrangements. This usually means providing a wide distribution of visible, clearly designated parking spots, close to the entrances of destinations being served.

Storage rooms or bicycle lockers may also be provided. In some cases large concentrations of bike parking may be more appropriate, sometimes being supervised and sometimes charging a fee. Examples include large bike parking facilities at public transport interchanges such as railway, subway, tram, bus stations, or at bike stations where they may be useful in Mixed-mode commuting.

Conversely, where cycling is seen as an unwelcome or inappropriate activity, or due to lack of knowledge about best practices, bicycle parking may simply not be provided or else placed at awkward, distant, and out-of-sight locations. Cyclists may be expressly forbidden from parking their bicycles at the most convenient locations. In April 2007, the authorities at the University of California's Santa Barbara campus started confiscating bicycles not parked at the allegedly inconvenient official bike racks. Often, property owners display signage on fencing to discourage bicyclists from locking their bicycles.

Pavement Engineering

Pavement engineering is a branch of civil engineering that uses engineering techniques to design and maintain flexible (asphalt) and rigid (concrete) pavements.

This includes streets and highways and involves knowledge of soils, hydraulics, and material properties. Pavement engineering involves new construction as well as rehabilitation and maintenance of existing pavements. Maintenance often involves using engineering judgment to make maintenance repairs with the highest long-term benefit and lowest cost. The Pavement Condition Index (PCI) is an example of an engineering approach applied to existing pavements. Another example is the use of a falling weight deflectometre (FWD) to non-destructively test existing pavements. Calculation of pavement layer strengths can be performed from the resulting deflection data.

Four on the Floor! (Strength vs. Stability)

There are two issues that will be emphasized throughout this course that are crucial to the understanding of Architectonics:

- Strength : the capacity of the individual elements, which together make up a structural system, to withstand the load that is applied to it.
- Stability: the capability of a structural system to transmit various loadings safely to the ground.

These two critical issues are experienced daily from the moment that an individual is born. A newborn baby cannot even hold its own head upright. The large mass of the head requires a support system that has sufficient strength to enable the head to maintain its stability. This steadily increases as the bones, muscles and tendons of the skeletal and muscular systems increase in strength. Eventually the extra support provided by the arm or hand is no longer needed. The first challenge posed by gravity is overcome.

Crawling on four points of support proves to be a very stabile situation for quite a long time. The "leap" to the unstable two point stance is the next development in our understanding of the influence of gravity. Again, the structural system must develop to the point that the individual elements of the system have acquired sufficient strength. The first steps are made: an action of supreme coordination of hundreds of elements that becomes second nature to homo sapiens. The list can be extrapolated to touch on many aspects of the human experience; riding tricycles and bicycles, jumping on trampolines, exercising on parallel bars, sliding on ice skates, sailing in a heavy wind, rocking a small boat, the list is endless. These are part of the human experience and each and every one rely on an inherent understanding of strength and stability.

How many times has a parent scolded a child to "put four on the floor!!!"? What the parent really means to say is, "if you do not put all of the legs of your chair on the ground, you are going to tip over!" Both strength and stability issues are addressed in this simple exclamation. Under normal conditions, the elements which make up the chair (its legs, bracing and seat) can easily resist the implied vertical loads. The strength of the individual elements of the chair have been designed for this type of static load. The seat (as a horizontal load-bearing element) must transfer its load through a connection to the legs (vertical load-bearing elements). Granted, some chairs will withstand a greater load than others, but they all resist the pull of gravity on the person sitting in them. If the legs cannot support the applied load they will fracture or break. These are examples of strength failure.

The stability of the system of elements depends upon the orientation of the chair in space. When it stands upright, on all four legs, it is a stable system. If it is on it's side, the chair might not be able to resist the loads for which it was designed. As it is tilted onto the back two legs, the structural system loses its equilibrium. At a certain point the chair as a system becomes unstable, fails and gravity pulls the supported load to the ground. This is a stability failure. In this type of failure, the individual elements retain their strength even as the system fails. The chair (system) could also have failed if the two supporting legs had experienced a strength failure (broken).

In each of these situations the chair, as a structural system, has reached the limit of its strength. As the saying goes, a chain (structural system) is only as strong as the weakest link (element)!

Any structural system can be studied in light of these two issues. For example, the column of the Greek temple shown above is an element that can experience a strength (crushing) failure, or a system (buckling) failure. It is/was part of a larger structural system.

The Role of Structural Failure

The concepts of element strength and system stability are learned at an early age. They often appear in children's stories that have been repeated for decades. One favourite in North America is the story of the "Three Little Pigs." Each pig desired to build a house to protect itself from the Big, Bad Wolf. One pig built a house of straw while another built a house of sticks and twigs. The third pig erected a house of brick. The story follows that the big, bad wolf came to the homes of the three pigs. He drew a huge breath and huffed and puffed until he blew the first house down. He chased the pig into the second house where he repeated his imitation of a typhoon. This house was also blown over and the pigs fled into the brick house. In front of the brick house the wolf hyperventilated. His grandest attempts to blow over this ediface failed and so he did not have pork for dinner. The story teaches children that a brick building is the strongest and most stable becuase it is the only one which could resist the huffing and puffing. The straw and stick buildings were blown away in the wind and only within the brick structure do the three pigs find their security. This lesson is correct if the load that is of greatest concern is a tornado, typhoon, hurricane or similar lateral load. However, the brick house would have been the worst structure to live in if it was built in an earthquake zone. In fact, an unreinforced masonry building is much more life threatening in an earthquake when compared to a straw hut. Then again, another response would be appropriate if the house was in a region where winds were not a real problem, but extreme temperature variations were critical. The same would be true if mobility was the most significant issue. Each of these are specific loading conditions that could lead to structural failure. It is critical that the appropriate loads and loading combinations are considered for each and every structure depending on use and location. Every one is different!

"Humpy Dumpty" is another childrens' tale with a structural lesson. He was a good egg who sat on a wall for quite a while. One day he lost his balance and fell to the ground where all of the king's men could not put Humpty back together again. The egg shell is a wonderfully efficient, naturally occurring thin structure that is extremly strong under evenly distributed loading conditions. One type of load which a thin shell cannot resist is a sharp point load. Thus, when Humpty fell to the ground he loaded his wonderful structure in a way that caused a strength and system failure. Building Codes are often considered to be the bane of architectural designers. However, they have been a part of the profession for thousands of years. The illustration above is a part of the Code of Hammurabi that was written about 2200BC. The Code can be translated as:

If a builder builds a house for a man and does not make its construction firm and the house collapses and causes the death of the owner of the house - that builder shall be put to death. If it causes the death of a son of the owner - a son of that builder shall be put to death.

If it causes the death of a slave of the owner - the builder shall give the owner a slave of equal value. If it destroys property - the builder shall restore whatever it destroyed and because the builder did not make the house firm, shall rebuild the house which collapsed at his own expense.

If a builder builds a house and does not make its construction meet the requirements and a wall falls in - that builder shall strengthen the wall at his own expense.

The negative incentive of the Code of Hammurabi does not bode well for innovation. However, it clearly demonstrates the point that building codes have existed for a long time, and are not likely to go away. Structural designers are often considered to be a very conservative group. One of the reasons for this could be found in such a building code. In addition, when a building experiences structural failure the results are usually quite different than if the air conditioning system fails. Although some aspects of the building codes seem ridiculous, they are in place to guarantee the public's safety and should be respected.

Each and every one of us is a structural designer in their own way. The act of walking down a set of stairs or pushing a wheelchair forward or swimming across a pool requires an understanding of gravity, the flow of forces and how to work with them.

Petroski notes: "The Concept of Failure is central to understanding engineering, it has as its first and foremost objective the obviation of failure. Design, even structural design, is a human endeavour and thus it is subject to error. Due to this, some designs are destined to fail. This can lead to a loss of life which in itself is tragic, but a deeper unforgivable tragedy exists when the lessons of the failure are understood and allowed to occur again."

What is a Force?

The one constant around the world is the action of gravity upon each and every structure that is erected. The primary function of all structural design is to make a building stand-up. Understanding architectonics will enable a designer to include these issues as part of a design language that will create a significantly clearer architectural expression. Primary to this study is the concept of a force. A force is actually a very abstract conception. It can be defined, but it cannot become physically apparent until it meets resistance. Imagine a six foot tall block of ice sliding along a frictionless surface laid inside of a hockey rink. If a person props herself against the wall and tries to stop the ice she will then perceive the force imparted by the block of ice. She transfers the force of the block of ice into a force that moves into the ground. Thus, if the ground could actually experience a push, it would as well as the slide is arrested.

A "force" is an action that changes, or tends to change, the state of motion of the body upon which it acts. It is a vector quantity that can can be represented either mathematically or graphically. A complete description of a force MUST include its:

1. Magnitude
2. Direction qnd Sense
3. Point of Application.

The Magnitude is most often expressed in the units of pounds (lbs), newtons (N), kilo-newtons (KN) or kilo-pounds (KIPS; 1 Kip = 1000 lbs). The magnitude is represented graphically by the scaled length of the arrow which represents the force. Graphic statics depends upon the accurate representation of the magnitude of each force acting upon a body.

The Direction of a force is discerned by oberserving the line of action of the vector of the force. The Sense refers to the direction of the movement of the vector along that line of action.

The sense is always represented on the vector by an arrowhead. These two attributes can be either a written verbal description or more conveniently expressed in terms of 360 degrees. In the later case, one begins with zero and increases clockwise with the direction of the arrowhead until 360 is reached.

The sense of the force can also be expressed as a positive or negative sign. This is especially useful when combining forces algebraically.

It is very important not to confuse the direction and sense. The direction always relates to the line of action of the vector, and the sense is the way in which the vector would move along that line. In the example, the sense of the 100 lb force would be "down and to the left" or "210 degrees."

The Point of Application is often overlooked in the description of a force. However, it is fully as important as the magnitude of the force!

It is the exact location of the application of a force on a body. It can either be a relative measurement or a set of coordinates.

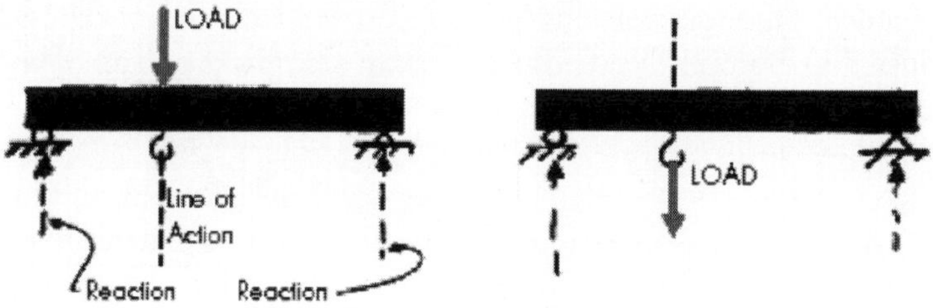

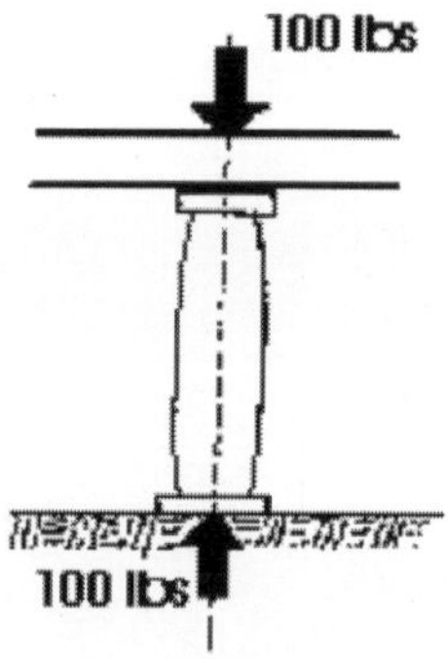

A 100 kip force is applied to the stone column in the diagram. The column will experience this force at every point along the line of action of the force. As a matter of fact, the force will also be transferred to the ground which is supporting the column. Thus, the earth below the column along the line of action of the force will also experience the 100 kip load. This illustrates the Principle of Transmissibility. The point of application of an external force acting on a body (structure) may be transmitted anywhere along the force's line of action without affecting the other external forces (reactions and loads) acting on that body. This means that there is NO NET CHANGE in the static effect upon any body if the body is in equilibrium. This can be illustrated with the following diagram.

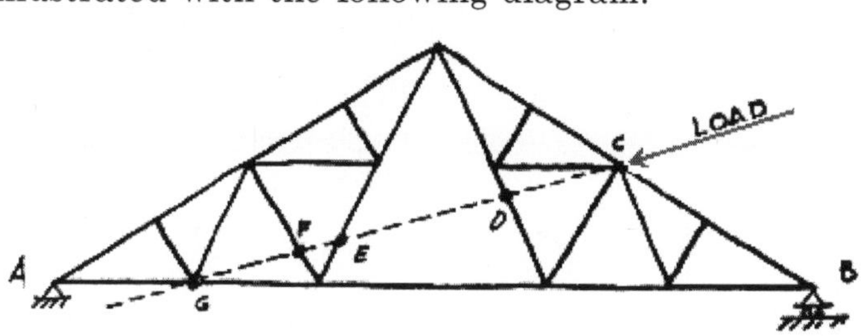

Assume that a beam is supported at its ends. A load is applied to the top of the beam that is acting downward. This load could be a person standing on the beam. The load creates reactions that push up at the two points of support. The line of action of the load (person) on the beam also passes through a hook that is attached to the underside of the beam. Now, if the person standing on top of the beam would climb down and hold on to the hook exactly below the point where they were previously located so that the lines of action were exactly the same, the reactions at the ends of the beams would not change. This is because the load of the person is still acting along the same line of action. As long as a load is applied at any point along the line of action the external reactions will not change.

The truss above is loaded with a force that is applied at point C. This load creates reactions at the two supports A and B. The load on the truss could move anywhere along the line of action and the external reactions at A and B would remain the same. That means that if the load was applied at points D, E, F or G the reactions at A and B would not change.

Note that the only point of discussion at this moment is the fact that the external reactions will not change. It is clear that the internal forces will vary greatly within the truss as the force is moved along the line of action. This issue will be discussed in a later lecture.

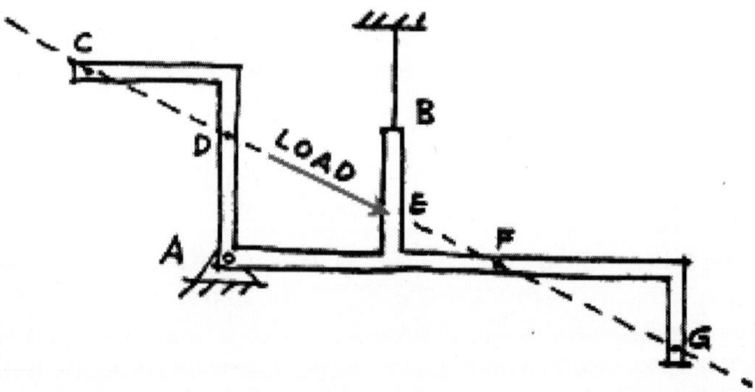

This strangely shaped beam is another example of a structure that is loaded at a specific point, namely E. In order for the structure to remain at rest there must be reactions of some kind at points A and B. The reaction at B is a tension reaction since it is a cable. Again, if the load was applied at points C, D, E, F or G on this rigid body (structure) the reactions at A and at B would remain exactly the same.

Only the external forces (reactions) remain unchanged. Some of the internal resisting forces within the elements of the structure change as the load is applied at different points along its line of action.

This illustrates one important difference between INTERNAL and EXTERNAL forces.

The Principle of Transmissibility applies to any body (blobs, balloons, simple beams, crooked beams, trusses, shells, etc.). It is independent of the body's size or shape.

Moments

The Moment of a force is a measure of its tendency to cause a body to rotate about a specific point or axis. This is different from the tendency for a body to move, or translate, in the direction of the force. In order for a moment to develop, the force must act upon the body in such a manner that the body would begin to twist. This occurs every time a force is applied so that it does not pass through the centroid of the body. A moment is due to a force not having an equal and opposite force directly along it's line of action.

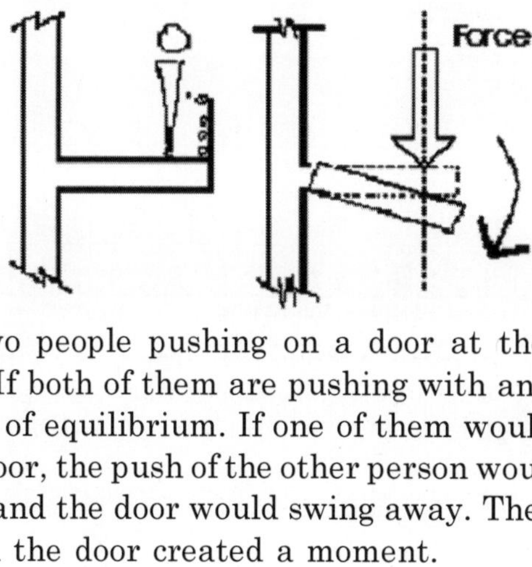

Imagine two people pushing on a door at the doorknob from opposite sides. If both of them are pushing with an equal force then there is a state of equilibrium. If one of them would suddenly jump back from the door, the push of the other person would no longer have any opposition and the door would swing away. The person who was still pushing on the door created a moment.

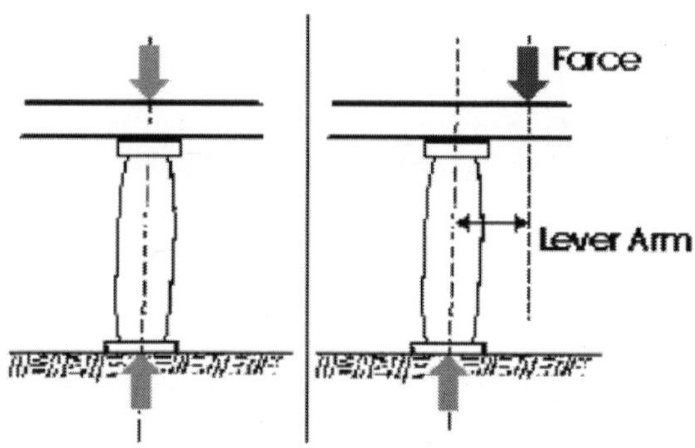

The magnitude of the moment of a force acting about a point or axis is directly proportional to the distance of the force from the point or axis. It is defined as the product of the force (F) and the moment arm (d). The moment arm or lever arm is the perpendicular distance between the line of action of the force and the centre of moments.

The Centre of Moments may be the actual point about which the force causes rotation. It may also be a reference point or axis about which the force may be considered as causing rotation. It does not matter as long as a specific point is always taken as the reference point. The latter case is much more common situation in structural design problems. A moment is expressed in units of foot-pounds, kip-feet, newton-metres, or kilonewton-metres. A moment also has a sense; it is either clockwise or counter-clockwise. The most common way to express a moment is :

$$Moment = Force \times Distance$$

$$M = F \times d$$

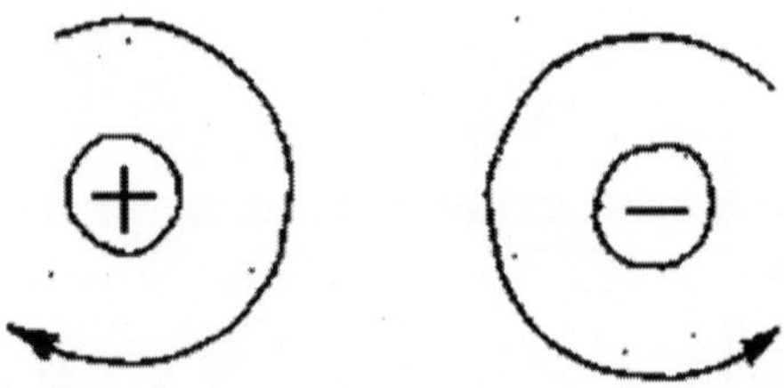

The example shown is a wrench on a nut (at C) that has a force applied to it. The force is applied at a distance of 12 inches from the nut. The centre of moments could be point C or points A or B. The moment arm for calculating the moment around point C is 12 inches. The magnitude of the moment of the 100 pound force about point C is 12 inches multiplied by the force of 100 pounds to give a moment of 1200 inch-pounds (or 100 foot-pounds). Similarly, the moment about point A can be found to be 800 inch-pounds.

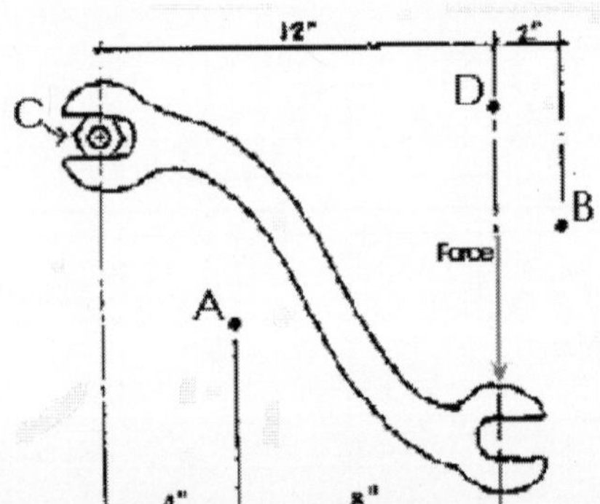

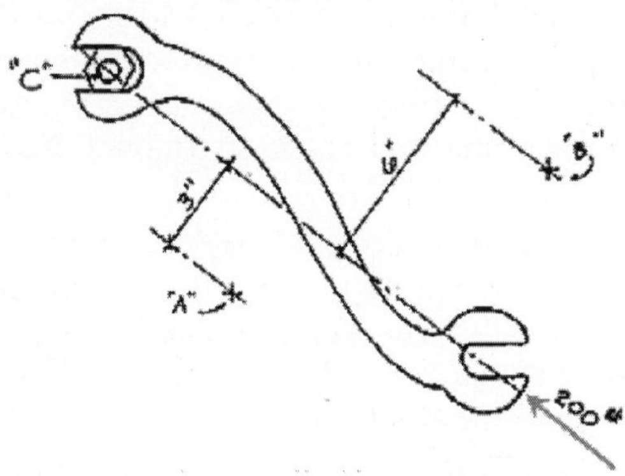

The direction of the rotation is important to understand in order to describe its effect on the body (structure). A moment will cause either a clockwise or counter-clockwise rotation about the centre of moments. It is essential that the direction of rotation about the centre of moments be understood. Since an international convention does not exist, in this course, a clockwise rotation about the centre of moments will be considered as positive; a counter-clockwise rotation about the centre of moments will be considered as negative. A moment causes a rotation about a point or axis. Thus, the moment of a force taken about any point that lies on its own line of action is zero. Such a force cannot cause a rotation because the moment arm is nonexistant.

In this second example, a 200 pound force is applied to the wrench. The moment of the 200 pound force applied at C is zero because 200 pounds x 0 inches = 0 inch-pounds (Fxd = M). In other words, there is no tendency for the 200 pound force to cause the wrench to rotate the nut. One could increase the magnitude of the force until the bolt finally broke off.

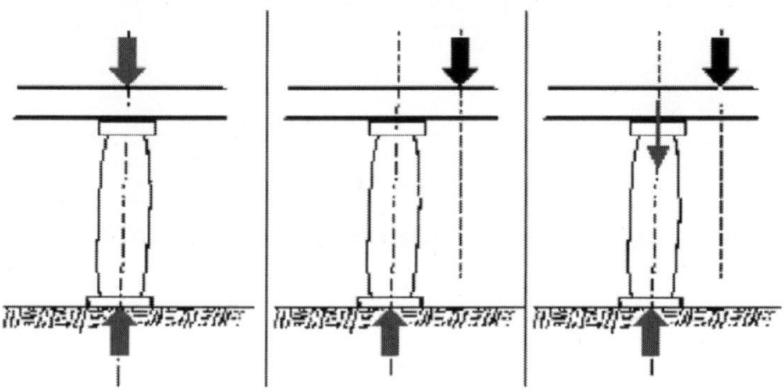

A moment can also be considered to be the result of forces detouring from a direct line drawn between the point of loading of a system and its supports. In this case, the blue force is an eccentric force. In order for it to reach the base of the column, it must make a detour through the beam. The greater the detour, the greater the moment. The most efficient structural systems have the least amount of detours possible. This will be discussed in more detail in Lecture 37 and later courses.

There are cases in which it is easier to calculate the moments of the componenets of a force around a certain point than it is to calculate the moment of the force itself. It could be that the determination of the perpendicular distance of the force is more difficult than determining the perpendicular distance of components of the force. The moment of several forces about a point is simply the algebraic sum of their component moments about the same point. When adding the moments of componenets, one must take great care to be consistent with the sense of each moment. It is often prudent to note the sense next to the moment when undertaking such problems.

Any difficulty with calculating a moment can usually be traced to one of the following:

1. The centre of moments has not been correctly established or clearly understood.
2. The assumed moment arm is not the PERPENDICULAR distance between the line of action of the force and the centre of moments.
3. The direction, or sense, of the rotation has been ignored or misunderstood.

Equilibrium

Architectural structures are normally stationary. Most clients, building officials and designers prefer that their structures remain static rather than move dynamically. There are specific loading conditions which are dynamic loads, but in each and every case a return to a stable and static state is desirable. Such a condition is known as equilibrium.

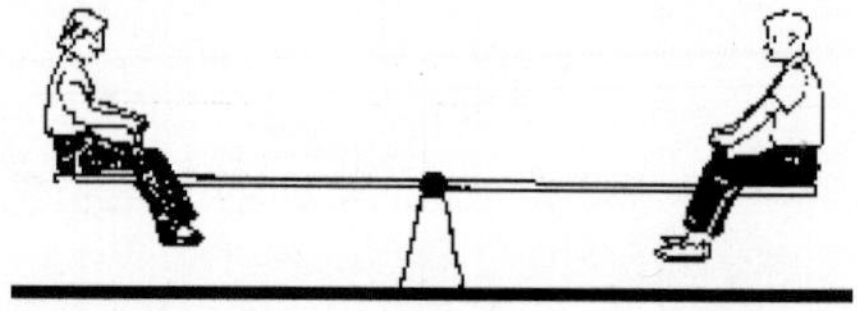

Various states of static equilibrium are experienced throughout one's life. Think of the "teeter-totter" at a playground or of a game of "tug-of-war." In the first case, two or more individuals sit upon a board which has been fixed to a fulcrum which allows rotation. If each of the individuals on the teeter totter weight exactly the same amount and sit at exactly the same distance from the fulcrum the teeter-totter will not move. A state of equilibrium has been achieved. The two will remain at rest until an action takes them out of equilibrium.

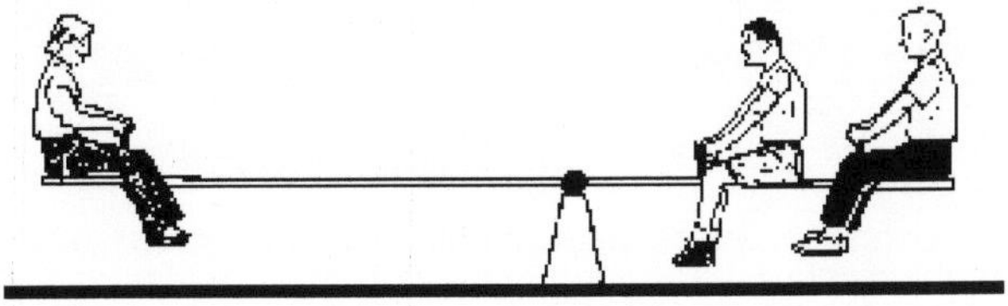

Such an action could be the addition of another person to the system or it could be that one of the original two would change their position slightly. In either case, the teeter-totter would most likely swing to one side and rest upon the ground. A new state of equilibrium would have been found. In this case, another boy climbed on behind the one already sitting on the right. In order to put the system back into equilibrium the girl on the left had to move backwards along the board until she was far away. The moment two boys created around the fulcrum doubled when the second boy climbed on. The girl knew that the only way that she could increase the magnitude of the moment she created would be to increase the moment arm. Thus she moved back until she was twice as far from the fulcrum. Now the system would be back in equilibrium.

Another example of a state of equilibrium is the game of "tug-of-war." A rope is pulled taught between two teams; each pulls with a force that equals the force of the opposite team. Assume in the figure that each team is pulling with a force of 220 pounds. As long as each team maintained a pull of 220 pounds the system is in equilibrium. If during this time a device would be inserted between the two teams to measure the magnitude of the tension force that the rope has anywhere along its length, it would read 220 pounds at each and every point. This would be true at ANY point along the rope. A structure is in equilibrium when all forces or moments acting upon it are balanced. This means that each and every force acting upon a body, or part of the body, is resisted by either another equal and opposite force or set of forces whose net result is zero. Issac Newton addressed this issue when he noted that a body is at rest will remain at rest until acted upon by an external force. Every structure that can be seen to remain standing on a daily basis is in equilibrium; it is at rest and each of its members, combination of its members or any part of a member that is supporting a load are also at rest. There is a net result of zero in all directions for all of the applied loads and reactions.

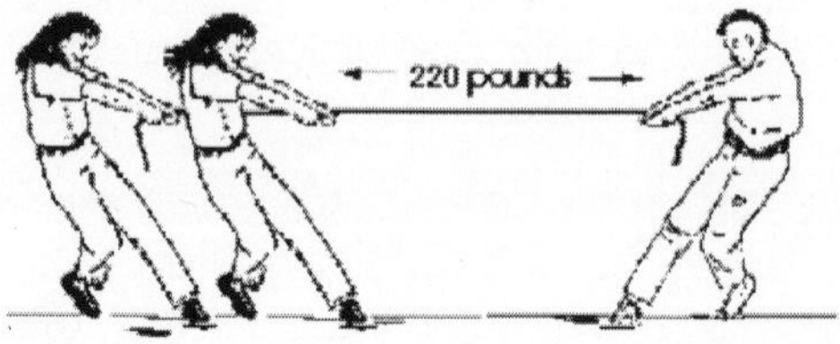

The beam of the top figure illustrates the system at rest under it's own weight. The lower figure shown another state in which a pair of scissors sits at one end and a thermos of coffee on the opposite side near the fulcrum. All of the forces and moments are balanced so that the system is in a stable equilibrium.

There are two types of equilibrium; External and Internal. External equilibrium encompasses the loads upon, and reactions of, a structural system as a whole. Internal equilibrium describes the various forces the are acting within every member of the system. There are conditions of equilibrium that must be satisfied for each case. These are:

Sum of All Vertical Forces (Fy) = 0

Sum of All Horizontal Forces (Fx) = 0

Sum of All Moments (Mz) = 0

(Sum of All Forces (Fz) = 0)

(Sum of All Moments (My) = 0)

(Sum of All Moments (Mx) = 0)

These six equations are all that can be used to determine every one of the forces that are acting with a structure. They are few, but very powerful. The first three are the most common equations and will be utilised in all of the problems associated with third course. The other three are only necessary when considering three-dimensional force systems.

4

Architectural Engineering

Architectural engineering, also known as building engineering, is the application of engineering principles and technology to building design and construction.

Figure: *César Pelli's Ratner Athletic Center uses cables, counterweights and masts as load-bearing devices.*

Definitions of an architectural engineer may refer to:

- An engineer in the structural, mechanical, electrical, construction or other engineering fields of building design and construction.
- A licensed engineering professional in parts of the United States.

- In informal contexts, and formally in some places, a professional synonymous with or similar to an architect. In some languages, "architect" is literally translated as "architectural engineer."

Engineering for Building

Structural engineering involves the analysis and design of physical objects such as buildings, bridges, equipment supports, towers and walls. Those concentrating on buildings are responsible for the structural performance of a large part of the built environment and are, sometimes, informally referred to as "building engineers". Structural engineers require expertise in strength of materials and in the seismic design of structures covered by earthquake engineering. Architectural Engineers sometimes practice structural as one aspect of their designs; the structural discipline when practised as a speciality works closely with architects and other engineering specialists.

Mechanical, Electrical, and Plumbing (MEP)

***Figure:** MEP room in a building*

Mechanical engineering and Electrical engineering are specialists, commonly referred to as "MEP" (mechanical, electrical, and plumbing) when engaged in the building design fields. Also known as "building services engineering" in the United Kingdom, Canada, and Australia. Mechanical engineers design and oversee the heating ventilation and air conditioning (HVAC), plumbing, and rain gutter systems. Plumbing designers often include design specifications for simple active fire protection systems, but for more complicated projects, fire protection engineers are often separately retained. Electrical engineers are responsible for the building's power distribution, telecommunication, fire alarm, signalisation, lightning protection and control systems, as well as lighting systems.

The Architectural Engineer (PE) in the United States

In many jurisdictions of the United States, the architectural engineer is a licensed engineering professional, usually a graduate of an architectural engineering university program preparing students to perform whole-building design in competition with architect-engineer teams; or for practice in one of structural, mechanical or electrical fields of building design, but with an appreciation of integrated architectural requirements. Formal architectural engineering education, following the engineering model of earlier disciplines, developed in the late 19th century, and became widespread in the United States by the mid 20th century. With the establishment of a specific "architectural engineering" NCEES Professional Engineering registration examination in the 1990's, and first offering in April 2003, architectural engineering became recognised as a distinct engineering discipline in the United States. Architectural engineers are not entitled to practice architecture unless they are also licensed as architects.

The Architect as Architectural Engineer

In some countries architecture, as a profession providing architectural services, is sometimes referred to as "architectural engineering". In others, such as in Japan, the terms "architecture" and "building engineering" are used synonymously. The practice of architecture includes the planning, designing and overseeing the building's construction.

In some languages, such as Korean and Arabic, "architect" is literally translated as "architectural engineer". In some countries, an "architectural engineer" (such as the *ingegnere edile* in Italy) is entitled to practice architecture and is often referred to as an architect. These individuals are often also structural engineers. In other countries, such as Germany, Austria, Pakistan and most of the Arabic countries, architecture graduates receive an engineering degree (*Dipl.-Ing. - Diplom-Ingenieur*). In Brazil, architects and engineers currently share the same accreditation process (CREA - Regional Council of Engineers and Architects). Besides traditional architecture design training, Brazilian architecture courses also offer complementary training in engineering disciplines such as structural, electrical, hydraulic and mechanical engineering. After graduation, architects can be fully responsible for most engineering design and construction, except highly specialised tasks such as road design and high voltage electrical.

Education

The architectural, structural, mechanical and electrical engineering branches each have well established educational requirements that are usually fulfilled by completion of a university program.

Architectural Engineering as a Single Integrated Field of Study

What differentiates architectural engineering as a separate and single, integrated field of study, compared to other engineering disciplines, is its multi-disciplined engineering approach. Through training in and appreciation of architecture, the field seeks integration of building systems within its overall building design. Architectural engineering includes the design of building systems including Heating, ventilation and air conditioning (HVAC), plumbing, fire protection, electrical, lighting, architectural acoustics, and structural systems. In some university programs, students are required to concentrate on one of the systems; in others, they can receive a generalist architectural or building engineering degree.

Facade Engineering

Building facades make a major contribution to the overall aesthetic and technical performance of a building. Facade engineers operate within technical divisions of facade manufacturing companies while some engineers provide facade consultancy for architects, building owners, cladding manufacturers and construction managers. Projects can include new buildings and recladding of existing buildings.

Facade engineers work in conjunction with other members of the design team including fire, acoustic and lighting engineers. Their goal is to balance the overall objectives of appearance, quality, performance, cost, durability and schedule. Facade engineering requires a blend of skills ranging from structural engineering through to building physics, architecture, manufacturing, materials science, dynamics, programming, procurement, project management and construction techniques.

Fire Protection Engineering

Fire Protection Engineering (also known as fire engineering or fire safety engineering) is the application of science and engineering principles to protect people and their environments from the destructive effects of fire and smoke.

The discipline of fire protection engineering includes, but is not exclusive to:

- Active fire protection - fire suppression systems, and fire alarm.
- Passive fire protection - fire and smoke barriers, space separation
- Smoke control and management
- Escape facilities- Emergency exits, Fire lifts etc.
- Building design, layout, and space planning
- Fire prevention programs
- Fire dynamics and fire modelling
- Human behaviour during fire events
- Risk analysis, including economic factors
- Wildfire Management.

Fire protection engineers identify risks and design safeguards that aid in preventing, controlling, and mitigating the effects of fires. Fire protection engineers assist architects, building owners and developers in evaluating buildings' life safety and property protection goals. FPEs are also employed as fire investigators, including such very large-scale cases as the analysis of the collapse of the World Trade Centres. NASA uses fire protection engineers in its space program to help improve safety. Fire protection engineers are also employed to provide 3rd party review for performance based fire engineering solutions submitted in support of local building regulation applications.

History

Fire protection engineering (FPE) roots dating as far back as Ancient Rome, when the Emperor Nero ordered the city to be rebuilt utilising passive fire protection methods, such as space separation and noncombustible building materials, after a catastrophic fire.The discipline of fire protection engineering emerged in the early 20th century as a distinct discipline, separate from civil, mechanical and chemical engineering, in response to new fire problems posed by the Industrial Revolution. Fire protection engineers of this era concerned themselves with devising methods to protect large factories, particularly spinning mills and other manufacturing properties. Another motivation to organize the discipline, define practices and conduct research to support innovations was in response to the catastrophic conflagrations and mass urban fires that swept many major cities during the latter half of the 19th century. The insurance industry also helped promote advancements in the fire protection engineering profession and the development of fire protection systems and equipment.

In 1903 the first degree program in fire protection engineering was initiated as the Armour Institute of Technology (later becoming part of the Illinois Institute of Technology).

As the 20th century emerged, several catastrophic firesresulted in changes to buildings codes to better protect people and property from fire. It was only in the latter half of the 20th Century that fire protection engineering emerged as a unique engineering profession. The primary reason for this emergence was the development of the "body of knowledge," specific to the profession that occurred after 1950. Other factors contributing to the growth of the profession include the start of the Institution of Fire Engineers in 1918 in the UK, and the Society of Fire Protection Engineers in 1950 in the USA, the emergence of independent fire protection consulting engineers, and the promulgation of engineering standards for fire protection.

Education

Fire protection engineers, like their counterparts in other engineering and scientific disciplines, undertake a formal course of education and continuing professional development to acquire and maintain their competence. This education typically includes foundation studies in mathematics, physics, chemistry, and technical writing. Professional engineering studies focus students on acquiring proficiency in material science, statics, dynamics, thermodynamics, fluid dynamics, heat transfer, engineering economics, ethics, Systems in engineering, reliability, and environmental psychology. Studies in combustion, probabilistic risk assessment or risk management, the design of fire suppression systems, the application and interpretation of model building codes, and the measurement and simulation of fire phenomena complete most curricula.

New Zealand was one of the first countries in the world to introduce Performance based assessment methods into their building codes in regard to fire safety. This occurred with the introduction of their 1991 Building Act. Professor Andy Buchanan, of the University of Canterbury, established the first post graduate and only course available in New Zealand, at the time, in fire safety engineering in 1995. Applicants to the course require a minimum qualification of a bachelors degree in engineering or bachelors degree in a limited list of science course. Notable alumni from the university of Canterbury include but is not limited to Sir Ernest Rutherford, Robert (Bob) Park, Roy Kerr, Michael P. Collins, John Britten and many others. A masters degree in fire engineering from the University of Canterbury is recognised under the Washington Accord.

In the United States, the University of Maryland (UMD) offers the only ABET-accredited B.S. degree program in Fire Protection Engineering, as well as graduate degrees and a distance M.Eng. program. Worcester Polytechnic Institute (WPI) offers a M.S. and a Ph.D. in Fire Protection Engineering. Oklahoma State University offers a ABET-accredited B.S. in Fire Protection and Safety Engineering Technology (established in 1937), the University of Houston–Downtown (UHD) is ABET-accredited for a B.S. in Fire Protection Engineering Technology, and the University of Cincinnati offers an associate degree in Fire Science and a bachelor's degree in Fire and Safety Engineering Technology as distance learning options, the only university in the U.S. and Canada to hold this distinction.. Other institutions, such as the University of New Haven, University of Kansas, Illinois Institute of Technology, University of California, Berkeley, Eastern Kentucky University, and the University of Texas at Austin have offered courses in Fire Protection Engineering or technology.

The practice of final fire sprinkler systems design and hydraulic calculations is commonly performed by design technicians who are often educated in-house at contracting firms throughout North America, with the objective of preparing designers for certification by testing by associations such as NICET (National Institute for Certification in Engineering Technologies). NICET certification is commonly used as a proof of competency for securing a license to design and install fire protection systems.

In Europe, the University of Edinburgh offers a degree in Fire Engineering and had its first fire research group in the 1970's. These activities are now conducted at the new BRE Centre for Fire Safety Engineering. Other European Universities active in the fire engineering are Luleå tekniska universitet, Lund University, Stord/Haugesund University College, University of Central Lancashire, University of Manchester, University of Ulster, University of Sheffield, University of Leeds, University of Greenwich, London South Bank University, University of Wales, Newport, Letterkenny Institute of Technology in Ireland and Otto-von-Guericke-Universität Magdeburg and Bergische Universität Wuppertal in Germany.

In Australia, Victoria University in Melbourne offers postgraduate courses in Building Fire Safety and Risk Engineering as does the University of Western Sydney. The Centre for Environmental Safety and Risk Engineering (CESARE) is a research unit under Victoria University and has facilities for research and testing of fire behaviour.

Professional Registration

Suitably qualified and experienced fire protection engineers may qualify for registration as a professional engineer. The recognition of fire protection engineering as a separate discipline varies from state to state in the United States. Few countries outside the United States regulate the professional practice of fire protection engineering as a discipline, although they may restrict the use of the title engineer in association with its practice. The titles *fire engineer* and *fire safety engineer* tend to be preferred outside the United States, especially in the United Kingdom and Commonwealth countries influenced by the British fire service. The Institution of Fire Engineers is one international organisation that qualifies many aspects of the training and qualifications of fire engineers.

Roof

A roof is the covering on the uppermost part of a building. A roof protects the building and its contents from the effects of weather. Structures that require roofs range from a letter box to a cathedral or stadium, dwellings being the most numerous. In most countries a roof protects primarily against rain. Depending upon the nature of the building, the roof may also protect against heat, sunlight, cold, snow and wind. Other types of structure, for example, a garden conservatory, might use roofing that protects against cold, wind and rain but admits light. A verandah may be roofed with material that protects against sunlight but admits the other elements. The characteristics of a roof are dependent upon the purpose of the building that it covers, the available roofing materials and the local traditions of construction and wider concepts of architectural design and practice and may also be governed by local or national legislation.

Design Elements

The elements in the design of a roof are:

- the material
- the construction
- the durability.

The material of a roof may range from banana leaves, wheaten straw or seagrass to lamininated glass, aluminium sheeting and precast concrete. In many parts of the world ceramic tiles have been the predominant roofing material for centuries.

Figure: *The roofs of Olomouc, Czech Republic*

Figure: *The roofs (with domes) of the Great Mosque of Kairouan also called the Mosque of Uqba, located in Kairouan, Tunisia*

The construction of a roof is determined by its method of support and how the underneath space is bridged and whether or not the roof is *pitched*. The *pitch* is the angle at which the roof rises from its lowest to highest point. Most domestic architecture, except in very dry regions, has roofs that are sloped, or *pitched*. The pitch is partly dependent upon stylistic factors, but has more to do with practicalities. Some types of roofing, for example thatch, require a steep pitch in order to be waterproof and durable. Other types of roofing, for example pantiles, are unstable on a steeply pitched roof but provide excellent weather protection at a relatively low angle. In regions where there is little rain, an almost flat roof with a slight runoff provides adequate protection against an occasional downpour.

The durability of a roof is a matter of concern because the roof is often the least accessible part of a building for purposes of repair and renewal, while its damage or destruction can have serious effects.

Form of a Roof

The shape of roofs differs greatly from region to region. The main factors which influence the shape of roofs are the climate and the materials available for roof structure and the outer covering.

The basic shapes of roofs are flat, skillion, gabled, hipped, arched and domed. There are many variations on these types. Roofs constructed of flat sections that are sloped are referred to as pitched roofs (generally if the angle exceeds 10 degrees). Pitched roofs, including gabled, hipped and skillion roofs, make up the greatest number of domestic roofs. Some roofs follow organic shapes, either by architectural design or because a flexible material such as that is used in the construction.

Parts of a Roof

There are two parts to a roof, its supporting structure and its outer skin, or uppermost weatherproof layer. In a minority of buildings, the outer layer is also a self-supporting structure. The roof structure is generally supported upon walls, although some building styles, for example, geodesic and A-frame, blur the distinction between wall and roof.

Support

The supporting structure of a roof usually comprises beams that are long and of strong, fairly rigid material such as timber, and since the mid 19th century, cast iron or steel. In countries that use bamboo extensively, the flexibility of the material causes a distinctive curving line to the roof, characteristic of Oriental architecture. Timber lends itself to a great variety of roof shapes. The timber structure can fulfil an aesthetic as well as practical function, when left exposed to view.

Stone lintels have been used to support roofs since prehistoric times, but cannot bridge large distances. The stone arch came into extensive use in the ancient Roman period and in variant forms could be used to span spaces up to 140 feet (43 m) across. The stone arch or vault, with or without ribs, dominated the roof structures of major architectural works for about 2,000 years, only giving way to iron beams with the Industrial Revolution and the designing of such buildings as Paxton's Crystal Palace, completed 1851.

With continual improvements in steel girders, these became the major structural support for large roofs, and eventually for ordinary houses as well. Another form of girder is the reinforced concrete beam, in which metal rods are encased in concrete, giving it greater strength under tension.

Outer Layer

This part of the roof shows great variation dependent upon availability of material. In simple vernacular architecture, roofing material is often vegetation, such as thatches, the most durable being sea grass with a life of perhaps 40 years. In many Asian countries bamboo is used both for the supporting structure and the outer layer where split bamboo stems are laid turned alternately and overlapped. In areas with an abundance of timber, wooden shingles are used, while in some countries the bark of certain trees can be peeled off in thick, heavy sheets and used for roofing.

The 20th century saw the manufacture of composition shingles which can last from a thin 20 year shingle to the thickest which are limited lifetime shingles, the cost depending on the thickness and durability of the shingle. When a layer of shingles wears out, they are usually stripped, along with the underlay and roofing nails, allowing a new layer to be installed. An alternative method is to install another layer directly over the worn layer. While this method is faster, it does not allow the roof sheathing to be inspected and water damage, often associated with worn shingles, to be repaired. Having multiple layers of old shingles under a new layer causes roofing nails to be located further from the sheathing, weakening their hold. The greatest concern with this method is that the weight of the extra material could exceed the dead load capacity of the roof structure and cause collapse.

Slate is an ideal, and durable material, while in the Swiss Alps roofs are made from huge slabs of stone, several inches thick. The slate roof is often considered the best type of roofing. A slate roof may last 75 to 150 years, and even longer. However, slate roofs are often expensive to install – in the USA, for example, a slate roof may have the same cost as the rest of the house. Often, the first part of a slate roof to fail is the fixing nails; they corrode, allowing the slates to slip. In the UK, this condition is known as "nail sickness". Because of this problem, fixing nails made of stainless steel or copper are recommended, and even these must be protected from the weather.

Asbestos, usually in bonded corrugated panels, has been used widely in the 20th century as an inexpensive, nonflammable roofing material with excellent insulating properties. Health and legal issues involved in the mining and handling of asbestos products means that it is no longer used as a new roofing material. However, many asbestos roofs continue to exist, particularly in South America and Asia. Roofs made of cut turf (modern ones known as Green roofs, traditional ones as sod roofs) have good insulating properties and are increasingly encouraged as a way of "greening" the Earth. Adobe roofs are roofs of clay, mixed with binding material such as straw or animal hair, and plastered on lathes to form a flat or gently sloped roof, usually in areas of low rainfall.

In areas where clay is plentiful, roofs of baked tiles have been the major form of roof. The casting and firing of roof tiles is an industry that is often associated with brickworks. While the shape and colour of tiles was once regionally distinctive, now tiles of many shapes and colours are produced commercially, to suit the taste and pocketbook of the purchaser.

Sheet metal in the form of copper and lead has also been used for many hundreds of years. Both are expensive but durable, the vast copper roof of Chartres Cathedral, oxidised to a pale green colour, having been in place for hundreds of years. Lead, which is sometimes used for church roofs, was most commonly used as flashing in valleys and around chimneys on domestic roofs, particularly those of slate. Copper was used for the same purpose.

In the 19th century, iron, electroplated with zinc to improve its resistance to rust, became a lightweight, easily-transported, waterproofing material. Its low cost and easy application made it the most accessible commercial roofing, world wide. Since then, many types of metal roofing have been developed. Steel shingle or standing-seam roofs last about 50 years or more depending on both the method of installation and the moisture barrier (underlayment) used and are between the cost of shingle roofs and slate roofs. In the 20th century a large number of roofing materials were developed, including roofs based on bitumen (already used in previous centuries), on rubber and on a range of synthetics such as thermoplastic and on fibreglass.

Tower

A tower is a tall structure, usually taller than it is wide, often by a significant margin. Towers are distinguished from masts by their lack of guy-wires.

Figure: *The Eiffel Tower in Paris, France.*

Towers are generally built to take advantage of their height, and can stand alone on the ground, or as part of a larger structure or device such as a fortified building or as an integral part of a bridge, the term also denoting a raised structure on a ship or other vehicle.

History

Towers have been used by mankind since prehistoric times. The oldest known may be the circular stone tower in walls of Neolithic Jericho (8000 BC). Some of the earliest towers were ziggurats, which existed in Sumerian architecture since the 4th millennium BC. The most famous ziggurats include the Sumerian Ziggurat of Ur, built the 3rd millennium BC, and the Etemenanki, one of the most famous examples of Babylonian architecture. The latter was built in Babylon during the 2nd millennium BC and was considered the tallest tower of the ancient world.

Some of the earliest surviving examples are the broch structures in northern Scotland, which are conical towerhouses. These and other examples from Phoenician and Roman cultures emphasized the use of a tower in fortification and sentinel roles. For example, watchtower elements are found at Mogador from the first millennium BC, derived from Phoenician or Carthaginian origins.

The Romans utilised octagonal towers as elements of Diocletian's Palace in Croatia, which monument dates to approximately 300 AD, while the Servian Walls (4th century BC) and the Aurelian Walls (3rd century AD) featured square ones. The Chinese used towers as integrated elements of the Great Wall of China in 210 BC during the Qin Dynasty. Towers were also an important element of castles.

Another well known tower is the Leaning Tower of Pisa in Pisa, Italy built from 1173 until 1372. The Himalayan Towers are stone towers located chiefly in Tibet built approximately 14th to 15th century.

Mechanics

Up to a certain height, a tower can be made with the supporting structure with parallel sides. However, above a certain height, the compressive load of the material is exceeded and the tower will fail. This can be avoided if the tower's support structure tapers up the building.

A second limit is that of buckling- the structure requires sufficient stiffness to avoid breaking under the loads it faces, especially those due to winds. Many very tall towers have their support structures at the periphery of the building, which greatly increases the overall stiffness.

A third limit is dynamic; a tower is subject to varying winds, vortex shedding, seismic disturbances etc. These are often dealt with a combination of simple strength and stiffness, as well as in some cases tuned mass dampers to damp out movements. Varying or tapering the outer aspect of the tower with height avoids vibrations due to vortex shedding occurring along the entire building simultaneously.

Functions

Skyscrapers: A modern type of tower, the skyscraper, uses less ground space as a ratio of total building interior square footage. Skyscrapers are often not classified as towers, although most have the same design and structure of towers. In the United Kingdom, tall domestic buildings are referred to as tower blocks. In the United States, the World Trade Centre had the nickname the Twin Towers, a name shared with the Petronas Twin Towers in Kuala Lumpur. A tower's has very deep foundation's.

Strategic Advantages

The tower throughout history has provided its users with an advantage in surveying defensive positions and obtaining a better view of the surrounding areas, including battlefields. They were installed on defensive walls, or rolled near a target. Today, strategic-use towers are still used at prisons, military camps, and defensive perimetres.

Potential Energy

By using gravity to move objects or substances downward, a tower can be used to store items or liquids like a storage silo or a water tower, or aim an object into the earth such as a drilling tower. Ski-jump ramps use the same idea, and in the absence of a natural mountain slope or hill, can be human-made.

Communication Enhancement

In history, simple towers like lighthouses, bell towers, clock towers, signal towers and minarets were used to communicate information over greater distances. In more recent years, radio masts and cell phone towers facilitate communication by expanding the range of the transmitter. The CN Tower in Toronto, Canada was built as a communications tower, with the capability to act as both a transmitter and repeater. Its design also incorporated features to make it a tourist attraction, including the world's highest observation deck at 147 stories.

Transportation Support

Towers can also be used to support bridges, and can reach heights that rival some of the tallest buildings above-water. Their use is most prevalent in suspension bridges and cable-stayed bridges. The use of the pylon, a simple tower structure, has also helped to build railroad bridges, mass-transit systems, and harbours. Control towers are used to give visibility to help direct aviation traffic.

Other Towers

- To access tall or high objects: launch tower, service tower, service structure, scaffold, tower wagon
- To access atmospheric conditions aloft: wind turbine, meteorological measurement tower, tower telescope, solar power station

- To lift high tension cables for electrical power distribution transmission tower
- To take advantage of the temperature gradient inherent in a height differential: cooling tower, chimney
- To protect from exposure: BREN Tower
- For industrial production: shot tower
- For surveying: Survey tower
- To drop objects: Drop tube (drop tower), bomb tower, diving platform
- To test height-intensive applications: elevator test tower
- To improve structural integrity: thyristor tower
- To mimic towers or provide height for training purposes: fire tower, parachute tower
- As art: Shukhov Tower
- For recreation: rock climbing tower
- As a symbol: Tower of Babel, The Tower (Tarot card), church tower

The term "tower" is also sometimes used to refer to firefighting equipment with an extremely tall ladder designed for use in firefighting/ rescue operations involving high-rise buildings.

Wind Engineering

Wind engineering analyses effects of wind in the natural and the built environment and studies the possible damage, inconvenience or benefits which may result from wind. In the field of structural engineering it includes strong winds, which may cause discomfort, as well as extreme winds, such as in a tornado, hurricane or heavy storm, which may cause widespread destruction. In the fields of wind energy and air pollution it also includes low and moderate winds as these are relevant to electricity production resp. dispersion of contaminants.

Wind engineering draws upon meteorology, fluid dynamics, mechanics, Geographic Information Systems and a number of specialist engineering disciplines including aerodynamics, and structural dynamics. The tools used include atmospheric modells, atmospheric boundary layer wind tunnels and computational fluid dynamics modells.

Wind engineering involves, among other topics:

- Wind impact on structures (buildings, bridges, towers).
- Wind comfort near buildings.

- Effects of wind on the ventilation system in a building.
- Wind climate for wind energy.
- Air pollution near buildings.

Wind engineering may be considered by structural engineers to be closely related to earthquake engineering and explosion protection.

History

Wind Engineering as a separate discipline can be traced to the UK in the 1960's, when informal meetings were held at the [National Physical Laboratory (United Kingdom) | [National Physical Laboratory]], the Building Research Establishment and elsewhere.

Wind Loads on Buildings

The design of buildings must account for wind loads, and these are affected by wind shear. For engineering purposes, a power law wind speed profile may be defined as follows:

$$v_z = v_g \cdot \left(\frac{z}{z_g}\right)^{\frac{1}{\alpha}}, 0 < z < z_g$$

where:

v_z = speed of the wind at height z

v_g = gradient wind at gradient height z_g

α = exponential coefficient.

Typically, buildings are designed to resist a strong wind with a very long return period, such as 50 years or more. The design wind speed is determined from historical records using Extreme value theory to predict future extreme wind speeds.

Wind Turbines

Wind turbines are affected by wind shear. Vertical wind-speed profiles result in different wind speeds at the blades nearest to the ground level compared to those at the top of blade travel, and this in turn affects the turbine operation. The wind gradient can create a large bending moment in the shaft of a two bladed turbine when the blades are vertical. The reduced wind gradient over water means shorter and less expensive wind turbine towers can be used in shallow seas.

For wind turbine engineering, wind speed variation with height is often approximated using a power law:

$$v_w(h) = v_{ref} \cdot \left(\frac{h}{h_{ref}} \right)^a$$

where:

$v_w(h)$ = velocity of the wind at height h [m/s]

v_{ref} = velocity of the wind at some reference height h_{ref} [m]

a = Hellman exponent (aka power law exponent or shear exponent) (~= 1/7 in neutral flow, but can be >1)

Significance

The knowledge of wind engineering is used to analyse and design all high rise buildings, cable suspension bridges and cable-stayed bridges, electricity transmission towers and telecommunication towers and all other types of towers and chimneys. The wind load is the dominant load in the analysis of many tall buildings. So wind engineering is essential for the analysis and design of tall buildings. Again, wind load is a dominant load in the analysis and design of all long-span cable bridges.

Surveying

Surveying or land surveying is the technique, profession, and science of accurately determining the terrestrial or three-dimensional position of points and the distances and angles between them. These points are usually on the surface of the Earth, and they are often used to establish land maps and boundaries for ownership or governmental purposes. To accomplish their objective, surveyors use elements of mathematics (geometry and trigonometry), physics, engineering and law.

An alternative definition, per the American Congress on Surveying and Mapping (ACSM), is the science and art of making all essential measurements to determine the relative position of points and/or physical and cultural details above, on, or beneath the surface of the Earth, and to depict them in a usable form, or to establish the position of points and/or details.

Furthermore, as alluded to above, a particular type of surveying known as "land surveying" (also per ACSM) is the detailed study or inspection, as by gathering information through observations, measurements in the field, questionnaires, or research of legal instruments, and data analysis in the support of planning, designing, and establishing of property boundaries. It involves the re-establishment of cadastral surveys and land boundaries based on documents of record and historical evidence, as well as certifying surveys (as required by statute or local ordinance) of subdivision plats/ maps, registered land surveys, judicial surveys, and space delineation. Land surveying can include associated services such as mapping and related data accumulation, construction layout surveys, precision measurements of length, angle, elevation, area, and volume, as well as horizontal and vertical control surveys, and the analysis and utilisation of land survey data. Surveying has been an essential element in the development of the human environment since the beginning of recorded history (about 5,000 years ago). It is required in the planning and execution of nearly every form of construction. Its most familiar modern uses are in the fields of transport, building and construction, communications, mapping, and the definition of legal boundaries for land ownership.

History of Surveying

Surveying techniques have existed throughout much of recorded history. In ancient Egypt, when the Nile River overflowed its banks and washed out farm boundaries, boundaries were re-established by a rope stretcher, or surveyor, through the application of simple geometry. The nearly perfect squareness and north-south orientation of the Great Pyramid of Giza, built c. 2700 BC, affirm the Egyptians' command of surveying.

A brief history of surveying:

- The Egyptian land register (3000 BC).
- A recent reassessment of Stonehenge (c. 2500 BC) suggests that the monument was set out by prehistoric surveyors using peg and rope geometry.
- The Groma surveying instrument originated in Mesopotamia (early 1st millennium BC).
- Under the Romans, land surveyors were established as a profession, and they established the basic measurements under which the Roman Empire was divided, such as a tax register of conquered lands (300 AD).
- The rise of the Caliphate led to extensive surveying throughout the Arab Empire. Arabic surveyors invented a variety of specialised instruments for surveying, including:
  - o Instruments for accurate levelling: A wooden board with a plumb line and two hooks, an equilateral triangle with a plumb line and two hooks, and a reed level.

  - A rotating alhidade, used for accurate alignment.
  - A surveying astrolabe, used for alignment, measuring angles, triangulation, finding the width of a river, and the distance between two points separated by an impassable obstruction.
- In England, The Domesday Book by William the Conqueror (1086)
  - covered all England
  - contained names of the land owners, area, land quality, and specific information of the area's content and inhabitants.
  - did not include maps showing exact locations.

In the 18th century in Europe triangulation was used to build a hierarchy of networks to allow point positioning within a country. Highest in the hierarchy were triangulation networks. These were densified into networks of traverses (polygons), into which local mapping surveying measurements, usually with measuring tape, corner prism and the familiar red and white poles, are tied. For example, in the late 1970's, a team from the Ordnance Survey of Great Britain, originally under General William Roy began the Principal Triangulation of Britain using the specially built Ramsden theodolite. Large scale surveys are known as geodetic surveys.

- Continental Europe's Cadastre was created in 1808
  - founded by Napoleon I (Bonaparte)
  - contained numbers of the parcels of land (or just land), land usage, names etc., and value of the land
  - 100 million parcels of land, triangle survey, measurable survey, map scale: 1:2500 and 1:1250
  - spread fast around Europe, but faced problems especially in Mediterranean countries, Balkan, and Eastern Europe due to cadastre upkeep costs and troubles.

A cadastre loses its value if register and maps are not constantly updated. Because of the fundamental value of land and real estate to the local and global economy, land surveying was one of the first professions to require Professional Licensure. In many jurisdictions, the land surveyors license was the first Professional Licensure issued by the state, province, or federal government.

Surveying Techniques

Historically, distances were measured using a variety of means, such as with chains having links of a known length, for instance a

Gunter's chain, or measuring tapes made of steel or invar. To measure horizontal distances, these chains or tapes were pulled taut according to temperature, to reduce sagging and slack. Additionally, attempts to hold the measuring instrument level would be made. In instances of measuring up a slope, the surveyor might have to "break" (break chain) the measurement- use an increment less than the total length of the chain.

Historically, horizontal angles were measured using a compass, which would provide a magnetic bearing, from which deflections could be measured. This type of instrument was later improved, with more carefully scribed discs providing better angular resolution, as well as through mounting telescopes with reticules for more-precise sighting atop the disc. Additionally, levels and calibrated circles allowing measurement of vertical angles were added, along with verniers for measurement to a fraction of a degree—such as with a turn-of-the-century transit.

The simplest method for measuring height is with an altimetre, basically a barometre, using air pressure as an indication of height. But surveying requires greater precision. A variety of means, such as precise levels (also known as differential levelling), have been developed to do this. With precise levelling, a series of measurements between two points are taken using an instrument and a measuring rod. Differentials in height between the measurements are added and subtracted in a series to derive the net difference in elevation between the two endpoints of the series. With the advent of the Global Positioning System (GPS), elevation can also be derived with sophisticated satellite receivers, but usually with somewhat less accuracy than with traditional precise levelling. However, the accuracies may be similar if the traditional levelling would have to be run over a long distance.

Triangulation is another method of horizontal location made almost obsolete by GPS. With the triangulation method, distances, elevations and directions between objects at great distance from one another can be determined. Since the early days of surveying, this was the primary method of determining accurate positions of objects for topographic maps of large areas. A surveyor first needs to know the horizontal distance between two of the objects. Then the height, distances and angular position of other objects can be derived, as long as they are visible from one of the original objects. High-accuracy transits or theodolites were used for this work, and angles between objects were measured repeatedly for increased accuracy.

Turning is a term used when referring to moving the level to take a elevation shot in a different location. When land surveying, there may be trees or other obstructions blocking the view from the level gun to the level rod. In order to "turn" the level gun, you mush first take a shot on the rod from your current location and record the elevation. Keeping the level rod in exactly the same location and elevation you may move the level gun to a different location where the level rod is still visible. Record the new elevation seen from the new location of the level rod and use the difference in elevations to find the new elevation of the level gun. Turning is not only used when there are obstructions in the way, but also when drastically changing elevations. You can turn up or down in elevation but the gun must always be at a higher elevation than the base of the rod. A level rod can usually be raised up to 25 feet high which enables the gun to be set much higher. However, if the gun is lower than the base of the rod, you will not be able to take a shot because the rod cannot be lowered beyond the ground elevation.

Surveying Equipment

As late as the 1990's, the basic tools used in planar surveying were a tape measure for determining shorter distances, a level to determine height or elevation differences, and a theodolite, set on a tripod, to measure angles (horizontal and vertical), combined with the process of triangulation. Starting from a position with known location and elevation, the distance and angles to the unknown point are measured.

A more modern instrument is a total station, which is a theodolite with an electronic distance measurement device (EDM). A total station can also be used for levelling when set to the horizontal plane. Since their introduction, total stations have made the technological shift from being optical-mechanical devices to being fully electronic.

Modern top-of-the-line total stations no longer require a reflector or prism (used to return the light pulses used for distancing) to return distance measurements, are fully robotic, and can even e-mail point data to the office computer and connect to satellite positioning systems, such as a Global Positioning System. Though real-time kinematic GPS systems have increased the speed of surveying, they are still horizontally accurate to only about 20 mm and vertically accurate to about 30–40 mm.

Total stations are still used widely, along with other types of surveying instruments. However, GPS systems do not work well in areas with dense tree cover or constructions. One-person robotic-guided total stations allow surveyors to gather precise measurements without extra workers to look through and turn the telescope or record data. A faster but expensive way to measure large areas (not details, and no obstacles) is with a helicopter, equipped with a laser scanner, combined with a GPS to determine the position and elevation of the helicopter. To increase precision, surveyors place beacons on the ground (about 20 km (12 mi) apart). This method reaches precisions between 5–40 cm (depending on flight height).

Types of Surveys and Applicability

- *ALTA/ACSM Survey*: a surveying standard jointly proposed by the American Land Title Association and the American Congress on Surveying and Mapping that incorporates elements of the boundary survey, mortgage survey, and topographic survey.
- *Archaeological survey*: used to accurately assess the relationship of archaeological sites in a landscape or to accurately record finds on an archaeological site.
- *As-built survey*: a survey carried out during or immediately after a construction project for record, completion evaluation and payment purposes. An as-built survey also known as a 'works as executed survey' documents the location of the recently constructed elements that are subject to completion evaluation. As built surveys are typically presented in red or redline and overlayed over existing design plans for direct comparison with design information.
- *Bathymetric survey:* a survey carried out to map the topography and features of the bed of an ocean, lake, river or other body of water.
- *Boundary survey*: a survey that establishes boundaries of a parcel using its legal description, which typically involves the setting or restoration of monuments or markers at the corners or along the lines of the parcel, often in the form of iron rods, pipes, or concrete monuments in the ground, or nails set in concrete or asphalt.
- *Deformation survey:* a survey to determine if a structure or object is changing shape or moving. The three-dimensional positions of specific points on an object are determined, a period of time is allowed to pass, these positions are then re-measured and calculated, and a comparison between the two sets of positions is made.

- *Engineering surveys*: those surveys associated with the engineering design (topographic, layout and as-built) often requiring geodetic computations beyond normal civil engineering practise.
- *Foundation survey*: a survey done to collect the positional data on a foundation that has been poured and is cured. This is done to ensure that the foundation was constructed in the location, and at the elevation, authorised in the *plot plan, site plan*, or *subdivision plan.*
- *Geological survey*: generic term for a survey conducted for the purpose of recording the geologically significant features of the area under investigation..
- *Hydrographic survey*: a survey conducted with the purpose of mapping the coastline and seabed for navigation, engineering, or resource management purposes.
- *Measured survey* : a building survey to produce plans of the building. Such a survey may be conducted before renovation works, for commercial purpose, or at end of the construction process "as built survey"
- *Mortgage survey or physical survey*: a simple survey that delineates land boundaries and building locations. In many places a mortgage survey is required by lending institutions as a precondition for a mortgage loan.
- *Soil survey*, or soil mapping, is the process of determining the soil types or other properties of the soil cover over a landscape, and mapping them for others to understand and use.
- *Structural survey*: a detailed inspection to report upon the physical condition and structural stability of a building or other structure and to highlight any work needed to maintain it in good repair.
- *Tape survey*: this type of survey is the most basic and inexpensive type of land survey. Popular in the middle part of the 20th century, tape surveys while being accurate for distance lack substantially in their accuracy of measuring angle and bearing. Standards that are practised by professional land surveyors.
- *Topographic survey*: a survey that measures the elevation of points on a particular piece of land, and presents them as contour lines on a plot.

Surveying as a Career

The basic principles of surveying have changed little over the ages, but the tools used by surveyors have evolved tremendously. Engineering, especially civil engineering, depends heavily on surveyors.

Whenever there are roads, railways, reservoir, dams, retaining walls, bridges or residential areas to be built, surveyors are involved. They establish the boundaries of legal descriptions and the boundaries of various lines of political divisions. They also provide advice and data for *geographical information systems* (GIS), computer database that contain data on land features and boundaries.

Surveyors must have a thorough knowledge of algebra, basic calculus, geometry, and trigonometry. They must also know the laws that deal with surveys, property, and contracts.

In addition, they must be able to use delicate instruments with accuracy and precision. In the United States, surveyors and civil engineers use units of feet wherein a survey foot is broken down into 10th and 100th. Many deed descriptions requiring distance calls are often expressed using these units (125.25 ft). On the subject of accuracy, surveyors are often held to a standard of one one-hundredth of a foot; about 1/8 inch. Calculation and mapping tolerances are much smaller wherein achieving near-perfect closures are desired. Though tolerances such as this will vary from project to project, in the field and day to day usage beyond a 100th of a foot is often impractical.

In most of the United States, surveying is recognised as a distinct profession apart from engineering. Licensing requirements vary by state, but they generally have components of education, experience and examinations. In the past, experience gained through an apprenticeship, together with passing a series of state-administered examinations, was required to attain licensure. Now, most states insist upon basic qualification of a degree in surveying, plus experience and examination requirements.

The licensing process typically follows two phases. First, upon graduation, the candidate may be eligible to take the Fundamentals of Land Surveying exam, to be certified upon passing and meeting all other requirements as a surveyor in training (SIT). Upon being certified as an SIT, the candidate then needs to gain additional experience to become eligible for the second phase. That typically consists of the Principles and Practice of Land Surveying exam along with a state-specific examination.

Licensed surveyors usually denote themselves with the letters P.S. (professional surveyor), L.S. (land surveyor), P.L.S. (professional land surveyor), R.L.S. (registered land surveyor), R.P.L.S. (Registered Professional Land Surveyor), or P.S.M. (professional surveyor and mapper) following their names, depending upon the dictates of their particular jurisdiction of registration.

In Canada, land Surveyors are registered to work in their respective province. The designation for a land surveyor breaks down by province, but follows the rule whereby the first letter indicates the province, followed by L.S. There is also a designation as a C.L.S. or Canada lands surveyor, who has the authority to work on Canada Lands, which include Indian Reserves, National Parks, the three territories and offshore lands. In many Commonwealth countries, the term Chartered Land Surveyor is used for someone holding a professional license to conduct surveys. A licensed land surveyor is typically required to sign and seal all plans, the format of which is dictated by their state jurisdiction, which shows their name and registration number. In many states, when setting boundary corners land surveyors are also required to place survey monuments bearing their registration numbers, typically in the form of capped iron rods, concrete monuments, or nails with washers.

Building Surveying

Building surveying emerged in the 1970's as a profession in the United Kingdom by a group of technically minded General Practice Surveyors. Building surveying is a recognised profession in Britain, Ireland, Australia and Hong Kong. In Australia in particular, due to risk mitigation and limitation factors, the employment of surveyors at all levels of the construction industry is widespread. There are still many countries where it is not widely recognised as a profession.

Building Surveyors are trained to some extent in all aspects of property but with specific training in Building Pathology, as such they have a wide understanding of the end implications of decisions taken by more specific professions and trades during the realisation process, thus making them ideal Project and Property Managers on the client side. Services that building surveyors undertake are broad but can include:

- Construction design and building works
- Project management and monitoring
- Property Legislation advice

- Insurance assessment and claims assistance
- Defect investigation and maintenance advice
- Building surveys and measured surveys
- Handling planning applications
- Building inspection to ensure compliance with building regulations
- Pre-acquisition surveys
- Negotiating dilapidations claims.

Building surveyors also advise on many aspects of construction including:

- design
- cost
- maintenance
- sustainability
- repair
- refurbishment
- restoration and preservation of buildings and monuments.

Clients of a building surveyor can be the government agencies, businesses and individuals. Surveyors work closely with architects, planners, quantity surveyors, engineers, homeowners and tenants groups. A building surveyor may be called to act as an expert witness. It is usual for building surveyors to earn a college degree before undertaking structured training to become a member of a professional organisation. With the enlargement of the European community, the profession of the building surveyor is becoming more widely known in other European states, particularly France, where many English-speaking people buy second homes.

Lidar Surveying – Three-dimensional laser scanning provides high definition surveying for architectural, as-built, and engineering surveys. Recent technological advances make it the most cost-effective and time-sensitive solution for providing the highest level of detail available for interior and exterior building work.

Land Surveyor

The job of the land surveyor is to find and mark certain locations on land. A typical location of interest, for example, is the boundary of a person's property. That boundary is described in legal documents and the land surveyor follows that description and locates the boundary on the physical land and marks it, so the owner knows what land he can legally use. As an example, such a legal description may refer to a point as being 120.25 feet south of some existing marker. The land surveyor in that case would find the existing marker and use measuring instruments to find the point 120.25 feet south of that, and place a new marker at that location. These markers are called monuments.

Over time, development, vandalism, and acts of nature often wreak havoc on monumentation, so the land surveyor is often forced to consider other evidence such as fence locations, woodlines, monuments on neighboring property, recollections of people, and other evidence.

Reference monumentation refers to actual physical points on the ground that define location of boundary lines that divide neighboring parcels as well as their respective corners. Also called survey control, they are most often 1/2" or 5/8" iron rebar rods or pipes placed at 18" minimum depth. These rods and/or pipes usually have an affixed plastic cap over the top bearing the responsible surveyors' name and license number. In addition to rods and pipes, surveyors often use 4x4" concrete posts at corners of large parcels or anywhere that would require more stability (e.g. beach sand). They place them three feet deep. In places where there is asphalt or concrete, it is common to place nails or aluminum alloy caps to re-establish boundary corners. Marks are meant to be durable, stable, and as "permanent" as possible. The aim is to provide sufficient marks so some marks will remain for future re-establishment of boundaries. The material and marking used on monuments placed to mark boundary corners are often subject to state laws. Cadastral land surveyors are licensed by governments. In the United States, cadastral surveys are typically conducted by the federal government, specifically through the Cadastral Surveys branch of the Bureau of Land Management (BLM), formerly the General Land Office (GLO). They consult with USFS, Park Service, Corps of Engineers, BIA, Fish and Wildlife Service, Bureau of Reclamation, etc. In states that have been organised per the Public Land Survey System (PLSS), surveyors carry out BLM Cadastral Surveys in accordance with that system.

A common use of a survey is to determine a legal property boundary. The first stage in such a survey is to research relevant title records such as deeds, survey monumentation (marks on the ground), and any public or private records that provide relevant data.

In order to properly establish the position for survey markers, the surveyor must then take measurements. To do this, the surveyor usually places a total station over various points on the ground and records distances taken with the EDM. The surveyor analyses the data and makes comparisons with existing records to determine evidence that can be used to establish boundary positions. The surveyor calculates the bearing and distance of lines between the boundary corners and total station positions and uses them to set out and mark the corners in the field. He may check measurements by measuring directly between places using a flexible tape.

The Art of Surveying

Many properties have considerable problems with regards to improper bounding, miscalculations in past surveys, titles, easements, and wildlife crossings. Also many properties are created from multiple divisions of a larger piece over the course of years, and with every additional division the risk of miscalculation increases. The result can be abutting properties not coinciding with adjacent parcels, resulting in hiatuses (gaps) and overlaps. The art plays a role when a surveyor must solve a puzzle using pieces that do not exactly fit together. In these cases, the solution is based upon the surveyor's research and interpretation, along with established procedures for resolving discrepancies.

5

Hydraulic Structure

A hydraulic structure is structure submerged or partially submerged in any body of water, which disrupts the natural flow of water. They can be used to divert, disrupt or completely stop the flow. An example of a hydraulic structure would be a dam, which slows the normal flow rate of the river in order to power turbines. A hydraulic structure can be built in rivers, a sea, or any body of water where there is a need for a change in the natural flow of water.

Hydraulic structures may also be used to measure the flow of water. When used to measure the flow of water, hydraulic structures are defined as a class of specially shaped, static devices over or through which water is directed in such a way that under free-flow conditions at a specified location (point of measurement) a known level to flow relationship exists. Hydraulic structures of this type can generally be divided into two categories: flumes and weirs.

Spillway

A spillway is a structure used to provide the controlled release of flows from a dam or levee into a downstream area, typically being the river that was dammed. In the UK they may be known as overflow channels. Spillways release floods so that the water does not overtop and damage or even destroy the dam. Except during flood periods, water does not normally flow over a spillway. In contrast, an *intake* is a structure used to release water on a regular basis for water supply, hydroelectricity generation, etc. Floodgates and fuse plugs may be designed into spillways to regulate water flow and dam height. Other uses of the term "spillway" include bypasses of dams or outlets of a channels used during highwater, and outlet channels carved through natural dams such as moraines.

Types

Figure: *Chute spillway of Llyn Brianne dam in Wales*

A spillway is located at the top of the reservoir pool. Dams may also have bottom outlets with valves or gates which may be operated to release flood flow, and a few dams lack overflow spillways and rely entirely on bottom outlets. There are two main types of spillways: controlled and uncontrolled.

A *controlled* spillway has mechanical structures or *gates* to regulate the rate of flow. This design allows nearly the full height of the dam to be used for water storage year-round, and flood waters can be released as required by opening one or more gates.

An *uncontrolled* spillway, in contrast, does not have gates; when the water rises above the lip or *crest* of the spillway it begins to be released from the reservoir. The rate of discharge is controlled only by the depth of water within the reservoir. All of the storage volume in the reservoir above the spillway crest can be used only for the temporary storage of floodwater, and cannot be used as water supply storage because it is normally empty.

In an intermediate type, normal level regulation of the reservoir is controlled by the mechanical gates. If inflow to the reservoir exceeds the gate's capacity, an artificial channel called either an *auxiliary* or *emergency* spillway that is blocked by a fuse plug dike will operate. The fuse plug is designed to over-top and wash out in case of a large flood, greater than the discharge capacity of the spillway gates. Although it may take many months to restore the fuse plug and channel after such an operation, the total damage and cost to repair is less than if the main water-retaining structures had been overtopped. The fuse plug concept is used where it would be very costly to build a spillway with capacity for the probable maximum flood.

Chute Spillways

Chute spillways are common and basic in design as they transfer excess water from behind the dam down a smooth decline into the river below. These are usually designed following an ogee curve. Most often, they are lined on the bottom and sides with concrete to protect the dam and topography. They may have a controlling device and some are thinner and multiply lined if space and funding are tight. In addition, they are not always intended to dissipate energy like stepped spillways. Chute spillways can be ingrained with a baffle of concrete blocks but usually have a 'flip lip' and/or dissipator basin which creates a hydraulic jump, protecting the toe of the dam from erosion.

Stepped Spillways

Figure: *A stepped chute baffled spillway of the Yeoman Hey Reservoir in England.*

Stepped channels and spillways have been used for over 3,000 years. Recently, new construction materials (e.g. RCC, gabions) and design techniques (e.g. embankment overtopping protection) have increased the interest in stepped spillways and chutes. The steps produce considerable energy dissipation along the chute and reduce the size of the required downstream energy dissipation basin. Research is still active on the topic, with newer developments on embankment dam overflow protection systems, converging spillways and small weir design.

Bell-Mouth Spillways

Some spillways are designed like an inverted bell so that water can enter all around the perimeter. These uncontrolled spillway devices are also called *morning glory, plughole, glory hole* or *bell-mouth* spillways. In areas where the surface of the reservoir may freeze, bell-mouth spillways are normally fitted with ice-breaking arrangements to prevent the spillway from becoming ice-bound. Chaffey Dam, located near Tamworth, New South Wales in Australia has a classic example of an inverted-bell spillway. In some cases bell-mouth spillways are gate controlled. The spillway at Hungry Horse Dam *(pictured right)*, the highest morning glory structure in the world, is controlled by a 64-by-12-foot (20 by 3.7 m) ring gate. However the largest remains in Lake Berryessa, measuring 72 feet in diameter at the lakes surface.

Other Types

Other spillway types include an *ogee crest* which over-tops a dam, a *side channel* that wraps around the topography of a dam and a *labyrinth* which uses a 'zig-zag' design to increase the sill length for a thinner design and increased discharge. There is also a *drop inlet* which resembles an intake for a hydroelectric power plant but transfers water from behind the dam directly through tunnels to the river downstream.

Design Considerations

The largest flood that needs be considered in the evaluation of a given project, regardless of whether a spillway is provided; i.e., a given project should have structures capable of safely passing the appropriate spillway design flood (SDF). A 100-year recurrence interval is the flood magnitude expected to be exceeded on the average of once in 100 years. It may also be expressed as an exceedance frequency with a one per cent chance of being exceeded in any given year.

Energy Dissipation

As water passes over a spillway and down the chute, potential energy converts into increasing kinetic energy. Failure to dissipate the water's energy can lead to scouring and erosion at the dam's toe (base). This can cause spillway damage and undermine the dam's stability. To put this energy in perspective, the spillways at Tarbela Dam could, at full capacity, produce 40,000 MW; about ten times the capacity of its power plant. The energy can be dissipated by addressing one or more parts of a spillway's design.

Steps: First, on the spillway surface itself by baffles and/or steps along the spillway.

Flip bucket: Second, at the base of a spillway, a *flip bucket* can create a hydraulic jump and deflect water upwards.

Ski jump: A *ski jump* can also direct water horizontally and eventually down into a plunge pool or two ski jumps can direct their water discharges to collide with one another.

Stilling basin: Third, a *stilling basin* at the terminus of a spillway serves to further dissipate energy and prevent erosion. They are usually filled with a relatively shallow depth of water and sometimes lined with concrete. A number of velocity-reducing components can be incorporated into the their design to include chute blocks, baffle blocks, wing walls, surface boils or an end sill.

Safety

Spillway gates may operate suddenly without warning, under remote control. Trespassers within the spillway run the risk of drowning. Spillways are usually fenced and equipped with locked gates to prevent casual trespassing within the structure. Warning signs, sirens, and other measures may be in place to warn users of the downstream area of sudden release of water. Operating protocols may require "cracking" a gate to release a small amount of water to warn persons downstream. The sudden closure of a spillway gate can result in the stranding of fish, and this is also usually avoided.

Dimensionless Momentum-Depth Relationship in Open-Channel Flow

Momentum in Open Channel Flow

What is Momentum: Momentum for one-dimensional flow in a channel can be given by the expression:

$$M = \frac{Q^2}{gA} + \bar{y}A$$

where:

- M is the momentum [ft^3]

- Q is the rate of flow [ft^3/s]
- g is the acceleration due to gravity [32.2 ft/s^2]
- A is the cross sectional area of flow [ft^2]
- $\bar{y}$ is the distance from the centroid of A to the water surface [ft]

For Open Channel Flow Calculations where momentum can be assumed to be conserved, we can equate the Momentum at an upstream location, M_1, to that at a downstream location, M_2, such that:

$$M_1 = M_2 = \frac{Q^2}{gA_1} + \bar{y}_1 A_1 = \frac{Q^2}{gA_2} + \bar{y}_2 A_2$$

Momentum in a Rectangular Channel

In the unique circumstance where the flow is in a rectangular channel (such as a laboratory flume), we can describe this relationship as Unit Momentum, by dividing both sides of the equation by the width of the channel. This produces $_uM$ in terms of ft^2, and is given by the equation:

$$uM_1 = uM_2 = \frac{q^2}{gy_1} + \frac{y_1^2}{2} = \frac{q^2}{gy_2} + \frac{y_2^2}{2}$$

where:

- $_uM$ is M/b
- q is Q/b
- b is the base width of the rectangular channel [ft]

Why is Momentum Important?

Momentum allows us to describe the characteristics of flow when energy is not conserved. HEC-RAS, a widely used computer model developed by the US Army Corps of Engineers for calculating water surface profiles, considers that any time flow passes through critical depth, the basic assumption of gradually varied flow required for the Energy Equation to be not applicable. Locations where flow may make such a transition include: significant changes in slope, channel geometry (e.g. bridge sections), grade control structures, and the confluence of water bodies. In these instances, HEC-RAS will use a form of the momentum equation to solve for the water surface elevation at an unknown location.

What are the characteristics of Momentum?

For discussion, we will consider an ideal, frictionless, rectangular channel. For each value of q, a unique curve can be produced where

M is shown as a function of depth. As is the case for specific energy, the minimum value of $_uM$, $_uM_{min}$, corresponds to critical depth. For each value of $_uM$ greater than $_uM_{min}$, there are two depths that can occur. These are called conjugate depths, and represent supercritical and subcritical alternatives for flow of a given $_uM$. Since hydraulic jumps conserve momentum, if the depth at the upstream or downstream end of a hydraulic jump is known, we can determine the unknown depth by drawing a vertical line through the known depth and reading its conjugate.

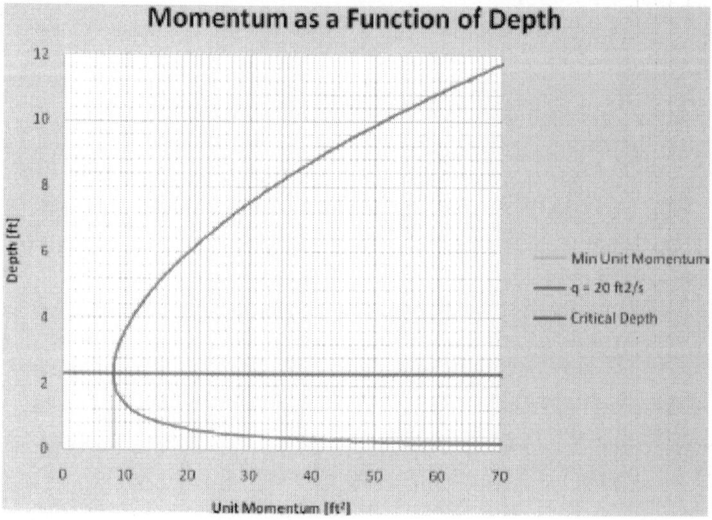

Dimensionless M'-y' Diagram

Why do we need a Dimensionless Momentum-Depth Relationship?:

Conjugate depths can be determined from curves like the one above. However, since this curve is unique for q = 20 ft²/s, we would have to develop a new curve for each rectangular channel of a given base width (or discharge). If we can establish a dimensionless relationship, we can apply the curve to any problem in which the cross-section is rectangular in shape. To create a dimensionless Momentum-Depth relationship, we will divide both sides by a normalizing value that will allow us to use a dimensionless relationship between Momentum and Depth for all values of q.

Derivation of the Dimensionless Momentum-Depth Relationship

Given that:

$$uM_1 = uM_2 = \frac{q^2}{gy_1} + \frac{y_1^2}{2} = \frac{q^2}{gy_2} + \frac{y_2^2}{2}$$

and that:

$$q^2 = gy_c^2$$

we can normalize the relationship between depth and Momentum by dividing both by the value of critical depth squared and substituting for q^2 to yield:

$$uMy_c^2 = \frac{gy_c^3}{gyy_c^2} + \frac{y^2}{2y_c^2}$$

where:

- y_c is the critical depth.

If we let M' = $_uM/y_c^2$, and y' = y/y_c, this equation becomes:

$$M' = \frac{1}{y'} + \frac{y'^2}{2}$$

The Dimensionless Momentum-Depth Diagram

By applying the conversion to dimensionless units described above, the Dimensionless Momentum-Depth diagram is produced below.

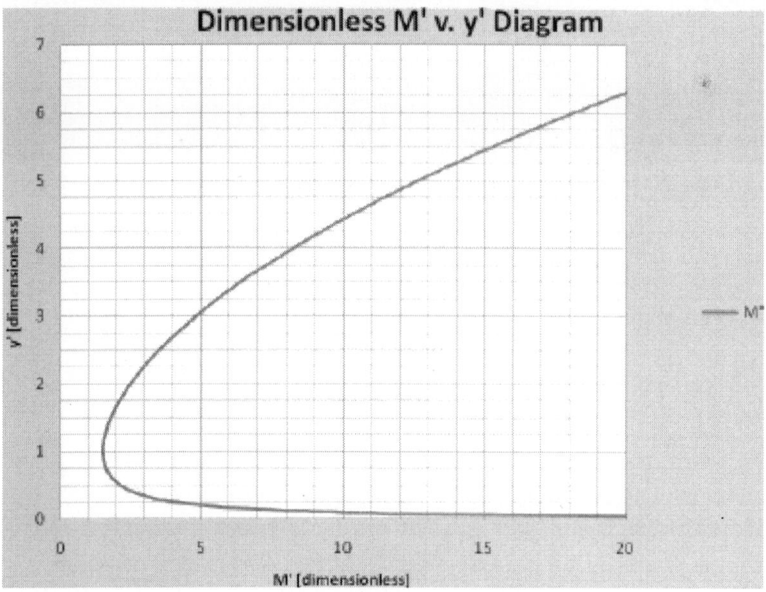

What is the Relationship between the Dimensionless Momentum-Depth Diagram and the Dimensionless Energy-Depth Diagram?

By close inspection of the Dimensionless Energy-Depth Diagram, an interesting conclusion can be drawn, which is that M' is the same function of y' as E' is of 1/y', and vice versa. This is demonstrated in the following chart that compares favourably to the chart of the Dimensionless Energy-Depth Diagram. Note that the only difference between the chart above and the one below is the values of the y-axis are the reciprocal of one another and that the scale has been changed to be consistent with the scale found in the discussion of Dimensionless Energy-Depth.

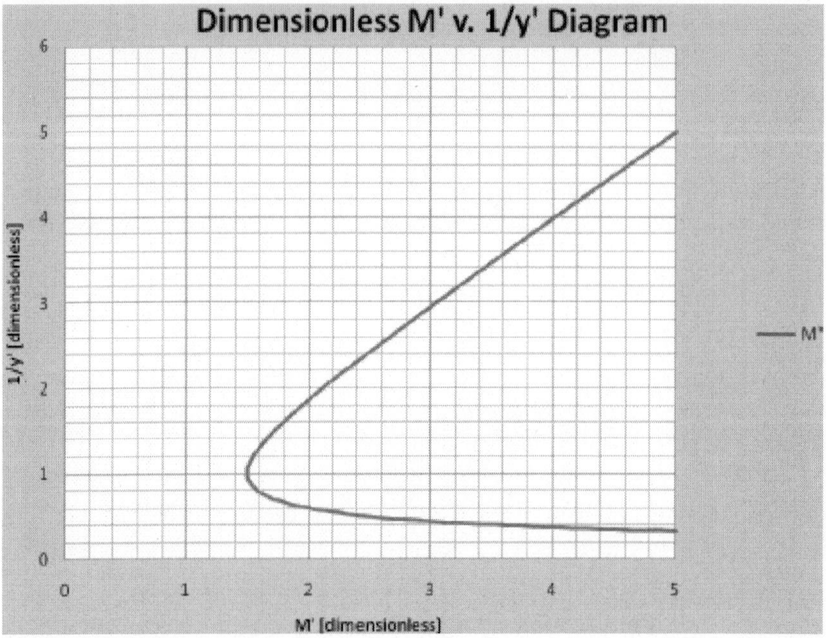

Because Energy and Momentum have this reciprocal relationship (found also in the non-dimensionless forms of these relationships), we can use a Dimensionless Energy-Depth Diagram to create a Dimensionless Momentum-Depth Diagram, and vice versa.

Solution of Simple Version of Hydraulic Jump with Dimensionless Diagram

To demonstrate the use of a Dimensionless Momentum-Depth Diagram in the solution of a simple hydraulic jump problem, let's consider a rectangular channel with a base width of 10 ft, and a flow rate of 100 ft^3/s, with a tailwater induced downstream depth of 6 ft. What is the depth of flow at the upstream end of the hydraulic jump?

Step 1 – Calculate q:

$$q = \frac{Q}{b}$$

$$q = \frac{100 ft^3 / s}{10 ft} = 10 ft^2 / s$$

Step 2 – Calculate y_c:

$$y_c = \sqrt[3]{\frac{q^2}{g}}$$

$$y_c = \sqrt[3]{\frac{(10 ft^2 / s)^2}{32.2 ft / s^2}} = 1.459 ft$$

(note-calculations are displayed to 3 decimal places to reduce rounding errors in Step 6)

Step 3 – Calculate y' for the downstream depth:

$$y' = \frac{y_d}{y_c}$$

$$y' = \frac{6 ft}{1.46 ft} = 4.112$$

Step 4 – Determine Conjugate Dimensionless Depth from Chart:

Using the Dimensionless Chart presented above, plot y' = 4.11 over to its intersection with the M' curve. Read down the chart to find the conjugate depth and determine the new y' from the left axis.

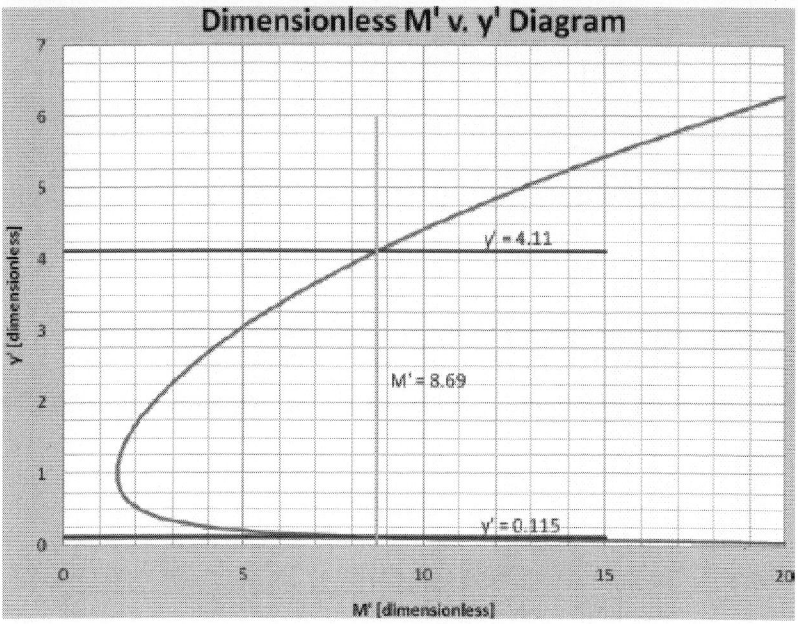

Step 5 – Calculate the upstream (conjugate) depth to 6 ft by converting y' = 0.115 to its actual depth:

$$y' = \frac{y_u}{y_c}$$

$$y_u = y'y_c = 0.115(1.459\,ft) = 0.168\,ft$$

Step 6 – Validation:

$$uM_u = \frac{q^2}{gy_1} + \frac{y_1^2}{2}$$

$$uM_u = \frac{10^2}{(32.2)0.168} + \frac{0.168^2}{2} = 18.500\,ft^2$$

and

$$uM_d = \frac{q^2}{gy_2} + \frac{y_2^2}{2}$$

$$uM_d = \frac{10^2}{(32.2)6} + \frac{6^2}{2} = 18.518\,ft^2$$

The difference between ${}_uM_u$ and ${}_uM_d$ is shown as 0.18 ft^2 due to rounding errors. Therefore, ${}_uM_u$ and ${}_uM_d$ are shown to represent the same unit momentum across the jump and momentum is conserved, validating the calculations using the Dimensionless Chart above.

This topic contribution was made in partial fulfillment of the requirements for Virginia Tech, Department of Civil and Environmental Engineering course: CEE 5984 – Open Channel Flow during the Fall 2010 semester.

Groundwater Flow Equation

Used in hydrogeology, the groundwater flow equation is the mathematical relationship which is used to describe the flow of groundwater through an aquifer. The transient flow of groundwater is described by a form of the diffusion equation, similar to that used in heat transfer to describe the flow of heat in a solid (heat conduction). The steady-state flow of groundwater is described by a form of the Laplace equation, which is a form of potential flow and has analogs in numerous fields.

The groundwater flow equation is often derived for a small representative elemental volume (REV), where the properties of the medium are assumed to be effectively constant. A mass balance is done on the water flowing in and out of this small volume, the flux terms in the relationship being expressed in terms of head by using the constituitive equation called Darcy's law, which requires that the flow is slow.

Mass Balance

A mass balance must be performed, and used along with Darcy's law, to arrive at the transient groundwater flow equation. This balance is analogous to the energy balance used in heat transfer to arrive at the heat equation. It is simply a statement of accounting, that for a given control volume, aside from sources or sinks, mass cannot be created or destroyed. The conservation of mass states that for a given increment of time (Δt) the difference between the mass flowing in across the boundaries, the mass flowing out across the boundaries, and the sources within the volume, is the change in storage.

$$\frac{\Delta M_{stor}}{\Delta t} = \frac{M_{in}}{\Delta t} - \frac{M_{out}}{\Delta t} - \frac{M_{gen}}{\Delta t}$$

Diffusion Equation (Transient Flow)

Mass can be represented as density times volume, and under most conditions, water can be considered incompressible (density does not depend on pressure). The mass fluxes across the boundaries then become volume fluxes (as are found in Darcy's law). Using Taylor series to represent the in and out flux terms across the boundaries of the control volume, and using the divergence theorem to turn the flux across the boundary into a flux over the entire volume, the final form of the groundwater flow equation (in differential form) is:

$$S_s \frac{\partial h}{\partial t} = -\nabla \cdot q - G.$$

This is known in other fields as the diffusion equation or heat equation, it is a parabolic partial differential equation (PDE). This mathematical statement indicates that the change in hydraulic head with time (left hand side) equals the negative divergence of the flux (q) and the source terms (G). This equation has both head and flux as unknowns, but Darcy's law relates flux to hydraulic heads, so substituting it in for the flux (q) leads to

$$S_s \frac{\partial h}{\partial t} = -\nabla \cdot (-k \nabla h) - G.$$

Now if hydraulic conductivity (k) is spatially uniform and isotropic (rather than a tensor), it can be taken out of the spatial derivative, simplifying them to the Laplacian, this makes the equation

$$S_s \frac{\partial h}{\partial t} = k\nabla^2 h - G.$$

Dividing through by the specific storage (S_s), puts hydraulic diffusivity ($\alpha = k/S_s$ or equivalently, $\alpha = T/S$) on the right hand side. The hydraulic diffusivity is proportional to the speed at which a finite pressure pulse will propagate through the system (large values of α lead to fast propagation of signals). The groundwater flow equation then becomes

$$\frac{\partial h}{\partial t} = \alpha\nabla^2 h - G.$$

Where the sink/source term, G, now has the same units but is divided by the appropriate storage term (as defined by the hydraulic diffusivity substitution).

Rectangular Cartesian Coordinates

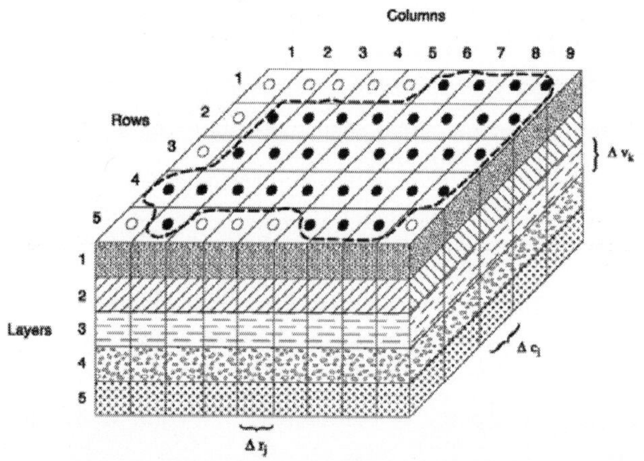

Figure: *Three-dimensional finite difference grid used in MODFLOW*

Especially when using rectangular grid finite-difference models (e.g. MODFLOW, made by the USGS), we deal with Cartesian coordinates. In these coordinates the general Laplacian operator becomes (for three-dimensional flow) specifically

$$\frac{\partial h}{\partial t} = \alpha\left[\frac{\partial^2 h}{\partial x^2} + \frac{\partial^2 h}{\partial y^2} + \frac{\partial^2 h}{\partial z^2}\right] - G.$$

MODFLOW code discretizes and simulates an orthogonal 3-D form of the governing groundwater flow equation. However, it has an option to run in a "quasi-3D" mode if the user wishes to do so; in this case the model deals with the vertically averaged T and S, rather than

k and S_s. In the quasi-3D mode, flow is calculated between 2D horizontal layers using the concept of leakage.

Circular Cylindrical Coordinates

Another useful coordinate system is 3D cylindrical coordinates (typically where a pumping well is a line source located at the origin — parallel to the z axis — causing converging radial flow). Under these conditions the above equation becomes (r being radial distance and θ being angle),

$$\frac{\partial h}{\partial t} = \alpha \left[\frac{\partial^2 h}{\partial r^2} + \frac{1}{r}\frac{\partial h}{\partial r} + \frac{1}{r^2}\frac{\partial^2 h}{\partial \theta^2} + \frac{\partial^2 h}{\partial z^2} \right] - G.$$

Assumptions

This equation represents flow to a pumping well (a sink of strength G), located at the origin. Both this equation and the Cartesian version above are the fundamental equation in groundwater flow, but to arrive at this point requires considerable simplification. Some of the main assumptions which went into both these equations are:

- the aquifer material is incompressible (no change in matrix due to changes in pressure — aka subsidence),
- the water is of constant density (incompressible),
- any external loads on the aquifer (e.g., overburden, atmospheric pressure) are constant,
- for the 1D radial problem the pumping well is fully penetrating a non-leaky aquifer,
- the groundwater is flowing slowly (Reynolds number less than unity), and
- the hydraulic conductivity (k) is an isotropic scalar.

Despite these large assumptions, the groundwater flow equation does a good job of representing the distribution of heads in aquifers due to a transient distribution of sources and sinks.

Laplace Equation (Steady-State Flow)

If the aquifer has recharging boundary conditions a steady-state may be reached (or it may be used as an approximation in many cases), and the diffusion equation (above) simplifies to the Laplace equation.

$$0 = \alpha \nabla^2 h$$

This equation states that hydraulic head is a harmonic function, and has many analogs in other fields. The Laplace equation can be solved using techniques, using similar assumptions stated above, but with the additional requirements of a steady-state flow field. A common method for solution of this equations in civil engineering and soil mechanics is to use the graphical technique of drawing flownets; where contour lines of hydraulic head and the stream function make a curvilinear grid, allowing complex geometries to be solved approximately. Steady-state flow to a pumping well (which never truly occurs, but is sometimes a useful approximation) is commonly called the Thiem solution.

Two-Dimensional Groundwater Flow

The above groundwater flow equations are valid for three dimensional flow. In unconfined aquifers, the solution to the 3D form of the equation is complicated by the presence of a free surface water table boundary condition: in addition to solving for the spatial distribution of heads, the location of this surface is also an unknown. This is a non-linear problem, even though the governing equation is linear.

An alternative formulation of the groundwater flow equation may be obtained by invoking the Dupuit assumption (or Dupuit-Forcheimer assumption), where it is assumed that heads do not vary in the vertical direction (i.e., $\partial h / \partial z = 0$). A horizontal water balance is applied to a long vertical column with area $\delta x \delta y$ extending from the aquifer base to the unsaturated surface. This distance is referred to as the saturated thickness, *b*. In a confined aquifer, the saturated thickness is determined by the height of the aquifer, *H*, and the pressure head is non-zero everywhere. In an unconfined aquifer, the saturated thickness is defined as the vertical distance between the water table surface and the aquifer base. If $\partial h / \partial z = 0$, and the aquifer base is at the zero datum, then the unconfined saturated thickness is equal to the head, i.e., $b=h$.

Assuming both the hydraulic conductivity and the horizontal components of flow are uniform along the entire saturated thickness of the aquifer (i.e., $\partial q_x / \partial z = 0$ and $\partial k / \partial z = 0$), we can express Darcy's law in terms of integrated discharges, Q_x and Q_y:

$$Q_x = \int_0^b q_x dz = -kb\frac{\partial h}{\partial x}$$

$$Q_y = \int_0^b q_y dz = -kb\frac{\partial h}{\partial y}$$

Inserting these into our mass balance expression, we obtain the general 2D governing equation for incompressible saturated groundwater flow:

$$\frac{\partial nb}{\partial t} = \nabla\cdot(kb\nabla h) + N.$$

Where n is the aquifer porosity. The source term, N (length per time), represents the addition of water in the vertical direction (e.g., recharge). By incorporating the correct definitions for saturated thickness, specific storage, and specific yield, we can transform this into two unique governing equations for confined and unconfined conditions:

$$S\frac{\partial h}{\partial t} = \nabla\cdot(kH\nabla h) + N.$$

(confined), where $S=S_s b$ is the aquifer storativity and

$$S_y\frac{\partial h}{\partial t} = \nabla\cdot(kh\nabla h) + N.$$

(unconfined), where S_y is the specific yield of the aquifer.

Note that the partial differential equation in the unconfined case is non-linear, whereas it is linear in the confined case. For unconfined steady-state flow, this non-linearity may be removed by expressing the PDE in terms of the head squared:

$$\nabla\cdot(k\nabla h^2) = -2N.$$

Or, for homogeneous aquifers,

$$\nabla^2 h^2 = -\frac{2N}{k}.$$

This formulation allows us to apply standard methods for solving linear PDEs in the case of unconfined flow. For heterogeneous aquifers with no recharge, Potential flow methods may be applied for mixed confined/unconfined cases.

Groundwater Energy Balance

The groundwater energy balance is the energy balance of a groundwater body in terms of incoming hydraulic energy associated with groundwater inflow into the body, energy associated with the outflow, energy conversion into heat due to friction of flow, and the resulting change of energy status and groundwater level.

Theory

When multiplying the horizontal velocity of groundwater (dimension, for example, m3/day per m^2 cross-sectional area) with the groundwater potential (dimension energy per m^3 water, or E/m^3) one obtains an energy flow (flux) in E/day per m^2 cross-sectional area.

Summation or integration of the energy flux in a vertical cross-section of unit width (say 1 m) from the lower flow boundary (the impermeable layer or base) up to the water table in an unconfined aquifer gives the energy flow *Fe* through the cross-section in *E*/day per m width of the aquifer.

While flowing, the groundwater loses energy due to friction of flow, i.e. hydraulic energy is converted into heat. At the same time, energy may be added with the recharge of water coming into the aquifer through the water table.

Thus one can make an hydraulic energy balance of a block of soil between two nearby cross-sections. In steady state, i.e. without change in energy status and without accumulation or depletion of water stored in the soil body, the energy flow in the first section plus the energy added by groundwater recharge between the sections minus the energy flow in the second section must equal the energy loss due to friction of flow.

In mathematical terms this balance can be obtained by differentiating the cross-sectional integral of *Fe* in the direction of flow using the Leibniz rule, taking into account that the level of the water table may change in the direction of flow. The mathematics is simplified using the Dupuit assumption.

The hydraulic friction losses can be described in analogy to *Joule's law* in electricity, where the friction losses are proportional to the square value of the current (flow) and the electrical resistance of the material through which the current occurs. In groundwater hydraulics (fluid dynamics, hydrodynamics) one often works with hydraulic conductivity (i.e. permeability of the soil for water), which is inversely proportional to the hydraulic resistance.

The resulting equation of the energy balance of groundwater flow can be used, for example, to calculate the shape of the water table between drains under specific aquifer conditions. For this a numerical solution can be used, taking small steps along the impermeable base.

The drainage equation is to be solved by trial and error (iterations), because the hydraulic potential is taken with respect to a reference level taken as the level of the water table at the water divide midway between the drains. When calculating the shape of the water table, its level at the water divide is initially not known. Therefore this level is to be assumed before the calculations on the shape of the water table can be started. According to the findings of the calculation procedure, the initial assumption is to be adjusted and the calculations are to be restarted until the level of the water table at the divide does not differ significantly from the assumed level.

The trial and error procedure is cumbersome and therefore computer programs may be developed to aid in the calculations.

Application

The energy balance of groundwater flow can be applied to flow of groundwater to subsurface drains. The computer program *EnDrain* compares the outcome of the traditional drain spacing equation, based on Darcy's law cum continuity equation (i.e. conservation of mass), with the solution obtained by the energy balance and it can be seen that drain spacings are wider in the latter case. This is owing to the introduction of the energy supplied by the incoming recharge.

6

Relative Density

Relative density, or specific gravity, is the ratio of the density (mass of a unit volume) of a substance to the density of a given reference material. Specific gravity usually means relative density with respect to water. The term "relative density" is often preferred in modern scientific usage.

If a substance's relative density is less than one then it is less dense than the reference; if greater than 1 then it is denser than the reference. If the relative density is exactly 1 then the densities are equal; that is, equal volumes of the two substances have the same mass. If the reference material is water then a substance with a relative density (or specific gravity) less than 1 will float in water. For example, an ice cube, with a relative density of about 0.91, will float. A substance with a relative density greater than 1 will sink.

Temperature and pressure must be specified for both the sample and the reference. Pressure is nearly always 1 atm equal to 101.325 kPa. Where it is not, it is more usual to specify the density directly. Temperatures for both sample and reference vary from industry to industry. In British brewing practice the specific gravity as specified above is multiplied by 1000. Specific gravity is commonly used in industry as a simple means of obtaining information about the concentration of solutions of various materials such as brines, sugar solutions (syrups, juices, honeys, brewers wort, must, etc.) and acids.

Basic Formulas

Relative density (*RD*) or specific gravity (*SG*) is a dimensionless quantity, as it is the ratio of either densities or weights

$$RD = \frac{\rho_{\text{substance}}}{\rho_{\text{reference}}}$$

where *RD* is relative density, $\rho_{substance}$ is the density of the substance being measured, and $\rho_{reference}$ is the density of the reference. (By convention ρ, the Greek letter rho, denotes density.)

The reference material can be indicated using subscripts: *RD*, which means "the relative density of *substance* with respect to *reference*". If the reference is not explicitly stated then it is normally assumed to be water at 4 °C (or, more precisely, 3.98 °C, which is the temperature at which water reaches its maximum density). In SI units, the density of water is (approximately) 1000 kg/m^3 or 1 g/cm^3, which makes relative density calculations particularly convenient: the density of the object only needs to be divided by 1000 or 1, depending on the units.

The relative density of gases is often measured with respect to dry air at a temperature of 20 °C and a pressure of 101.325 kPa absolute, which has a density of 1.205 kg/m^3. Relative density with respect to air can be obtained by

$$RD = \frac{\rho_{\text{gas}}}{\rho_{\text{air}}} \approx \frac{M_{\text{gas}}}{M_{\text{air}}}$$

Where *M* is the molar mass and the approximately equal sign is used because equality pertains only if 1 mol of the gas and 1 mol of air occupy the same volume at a given temperature and pressure i.e. they are both Ideal gases. Ideal behaviour is usually only seen at very low pressure. For example, one mol of an ideal gas occupies 22.414 L at 0 °C and 1 atmosphere whereas carbon dioxide has a molar volume of 22.259 L under those same conditions.

Temperature Dependence

The density of substances varies with temperature and pressure so that it is necessary to specify the temperatures and pressures at which the densities or weights were determined. It is nearly always the case that measurements are made at nominally 1 atmosphere (101.325 kPa the variations caused by changing weather patterns) but as specific gravity usually refers to highly incompressible aqueous solutions or other incompressible substances (such as petroleum products) variations in density caused by pressure are usually neglected at least where apparent specific gravity is being measured. For true (*in vacuo*) specific gravity calculations air pressure must be considered.

Temperatures are specified by the notation T_s/T_r) with T_s representing the temperature at which the sample's density was determined and T_r the temperature at which the reference (water) density is specified. For example SG (20°C/4°C) would be understood to mean that the density of the sample was determined at 20 °C and of the water at 4 °C. Taking into account different sample and reference temperatures we note that while SG_{H2O} = 1.000000 (20°C/20°C) it is also the case that SG_{H2O} = 0.998203/0.998840 = 0.998363 (20°C/4°C). Here temperature is being specified using the current ITS-90 scale and the densities used here and in the rest of this article are based on that scale. On the previous IPTS-68 scale the densities at 20 °C and 4 °C are, respectively, 0.9982071 and 0.9999720 resulting in an SG (20°C/4°C) value for water of 0.9982343.

The temperatures of the two materials may be explicitly stated in the density symbols; for example: relative density: $8.15_{4^\circ C}^{20^\circ C}$ or specific gravity: 2.432_0^{15}

where the superscript indicates the temperature at which the density of the material is measured, and the subscript indicates the temperature of the reference substance to which it is compared.

Uses

Relative density can also help quantify the buoyancy of a substance in a fluid, or determine the density of an unknown substance from the known density of another. Relative density is often used by geologists and mineralogists to help determine the mineral content of a rock or other sample. Gemologists use it as an aid in the identification of gemstones. Water is preferred as the reference because measurements are then easy to carry out in the field.

As the principal use of specific gravity measurements in industry is determination of the concentrations of substances in aqueous solutions and these are found in tables of SG vs concentration it is extremely important that the analyst enter the table with the correct form of specific gravity. For example, in the brewing industry, the Plato table, which lists sucrose concentration by weight against true SG, were originally (20 °C/4 °C) that is based on measurements of the density of sucrose solutions made at laboratory temperature (20 °C) but referenced to the density of water at 4 °C which is very close to the temperature at which water has its maximum density of $\rho(H_2O)$ equal to 0.999972 g/cm^3 (or 62.43 $lb_m.ft^{-3}$). The ASBC table in use today in North America, while it is derived from the original Plato table is for apparent specific gravity measurements at (20 °C/20 °C) on the IPTS-68 scale where the density of water is 0.9982071 g/cm^3. In the sugar, soft drink, honey, fruit juice and related industries sucrose concentration by weight is taken from this work which uses SG (17.5 °C/17.5 °C). As a final example, the British SG units are based on reference and sample temperatures of 60°F and are thus (15.56°C/15.56°C).

Measurement

Relative density can be calculated directly by measuring the density of a sample and dividing it by the (known) density of the reference substance. The density of the sample is simply its mass divided by its volume. Although mass is easy to measure, the volume of an irregularly shaped sample can be more difficult to ascertain. One method is to put the sample in a water-filled graduated cylinder and read off how much water it displaces. Alternatively the container can be filled to the brim, the sample immersed, and the volume of overflow measured. The surface tension of the water may keep a significant amount of water from overflowing, which is especially problematic for small samples. For this reason it is desirable to use a water container with as small a mouth as possible.

For each substance, the density, ρ, is given by

$$\rho = \frac{\text{Mass}}{\text{Volume}} = \frac{\text{Deflection} \times \dfrac{\text{Spring Constant}}{\text{Gravity}}}{\text{Displacement}_{\text{WaterLine}} \times \text{Area}_{\text{Cylinder}}}$$

When these densities are divided, references to the spring constant, gravity and cross-sectional area simply cancel, leaving

$$RD = \frac{\rho_{\text{object}}}{\rho_{\text{ref}}} = \frac{\dfrac{\text{Deflection}_{\text{Obj.}}}{\text{Displacement}_{\text{Obj.}}}}{\dfrac{\text{Deflection}_{\text{Ref.}}}{\text{Displacement}_{\text{Ref.}}}} = \frac{\dfrac{3\text{ in}}{20\text{ mm}}}{\dfrac{5\text{ in}}{34\text{ mm}}} = \frac{3\text{ in} \times 34\text{ mm}}{5\text{ in} \times 20\text{ mm}} = 1.02$$

Hydrostatic Weighing

Relative density is more easily and perhaps more accurately measured without measuring volume. Using a spring scale, the sample is weighed first in air and then in water. Relative density (with respect to water) can then be calculated using the following formula:

$$RD = \frac{W_{\text{air}}}{W_{\text{air}} - W_{\text{water}}}$$

where

W_{air} is the weight of the sample in air (measured in pounds-force, newtons, or some other unit of force)

W_{water} is the weight of the sample in water (measured in the same units).

This technique cannot easily be used to measure relative densities less than one, because the sample will then float. W_{water} becomes a negative quantity, representing the force needed to keep the sample underwater. Another practical method uses three measurements. The sample is weighed dry. Then a container filled to the brim with water is weighed, and weighed again with the sample immersed, after the displaced water has overflowed and been removed. Subtracting the last reading from the sum of the first two readings gives the weight of the displaced water. The relative density result is the dry sample weight divided by that of the displaced water. This method works with scales that can't easily accommodate a suspended sample, and also allows for measurement of samples that are less dense than water.

Hydrometer

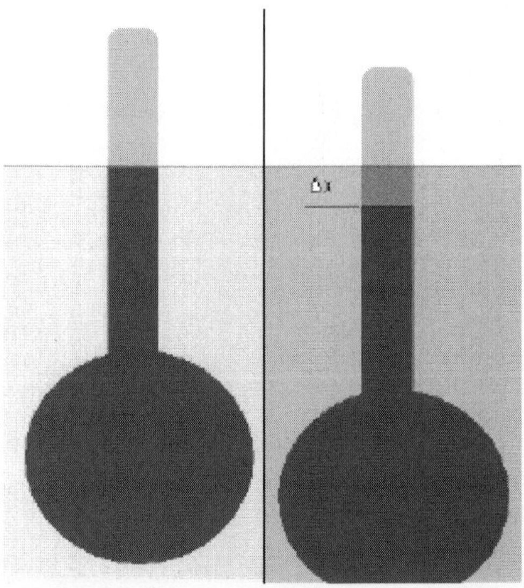

The relative density of a liquid can be measured using a hydrometre. This consists of a bulb attached to a stalk of constant cross-sectional area, as shown in the diagram to the right.

First the hydrometre is floated in the reference liquid (shown in light blue), and the displacement (the level of the liquid on the stalk) is marked (blue line). The reference could be any liquid, but in practice it is usually water.

The hydrometre is then floated in a liquid of unknown density (shown in green). The change in displacement, Δx, is noted. In the example depicted, the hydrometre has dropped slightly in the green liquid; hence its density is lower than that of the reference liquid. It is, of course, necessary that the hydrometre floats in both liquids.

The application of simple physical principles allows the relative density of the unknown liquid to be calculated from the change in displacement. (In practice the stalk of the hydrometre is pre-marked with graduations to facilitate this measurement.)

In the explanation that follows,

ρ_{ref} is the known density (mass per unit volume) of the reference liquid (typically water).

ρ_{new} *is the unknown density of the new (green) liquid.*

$RD_{\text{new/ref}}$ is the relative density of the new liquid with respect to the reference.

V is the volume of reference liquid displaced, i.e. the red volume in the diagram.

m is the mass of the entire hydrometre.

g is the local gravitational constant.

Δx is the change in displacement. In accordance with the way in which hydrometres are usually graduated, Δx is here taken to be negative if the displacement line rises on the stalk of the hydrometre, and positive if it falls. In the example depicted, Δx is negative.

A is the cross sectional area of the shaft.

Since the floating hydrometre is in static equilibrium, the downward gravitational force acting upon it must exactly balance the upward buoyancy force. The gravitational force acting on the hydrometre is simply its weight, mg. From the Archimedes buoyancy principle, the buoyancy force acting on the hydrometre is equal to the weight of liquid displaced. This weight is equal to the mass of liquid displaced multiplied by g, which in the case of the reference liquid is $\rho_{ref} Vg$. Setting these equal, we have

$$mg = \rho_{\text{ref}} Vg$$

or just

$$m = \rho_{\text{ref}} V \tag{1}$$

Exactly the same equation applies when the hydrometre is floating in the liquid being measured, except that the new volume is $V - A\Delta x$. Thus,

$$m = \rho_{new}(V - A\Delta x) \quad (2)$$

Combining (1) and (2) yields

$$RD_{new/ref} = \frac{\rho_{new}}{\rho_{ref}} = \frac{V}{V - A\Delta x} \quad (3)$$

But from (1) we have $V = m/\rho_{ref}$. Substituting into (3) gives

$$RD_{new/ref} = \frac{1}{1 - \frac{A\Delta x}{m}\rho_{ref}} \quad (4)$$

This equation allows the relative density to be calculated from the change in displacement, the known density of the reference liquid, and the known properties of the hydrometre. If Δx is small then, as a first-order approximation of the geometric series equation (4) can be written as:

$$RD_{new/ref} \approx 1 + \frac{A\Delta x}{m}\rho_{ref}$$

This shows that, for small Δx, changes in displacement are approximately proportional to changes in relative density.

Pycnometre

A pycnometre (from Greek: πõêvüò (puknos) meaning "dense"), also called pyknometre or specific gravity bottle, is a device used to determine the density of a liquid. A pycnometre is usually made of glass, with a close-fitting ground glass stopper with a capillary tube through it, so that air bubbles may escape from the apparatus. This device enables a liquid's density to be measured accurately by reference to an appropriate working fluid, such as water or mercury, using an analytical balance.

If the flask is weighed empty, full of water, and full of a liquid whose specific gravity is desired, the specific gravity of the liquid can easily be calculated. The particle density of a powder, to which the usual method of weighing cannot be applied, can also be determined with a pycnometre. The powder is added to the pycnometre, which is then weighed, giving the weight of the powder sample. The pycnometre is then filled with a liquid of known density, in which the powder is completely insoluble. The weight of the displaced liquid can then be determined, and hence the specific gravity of the powder.

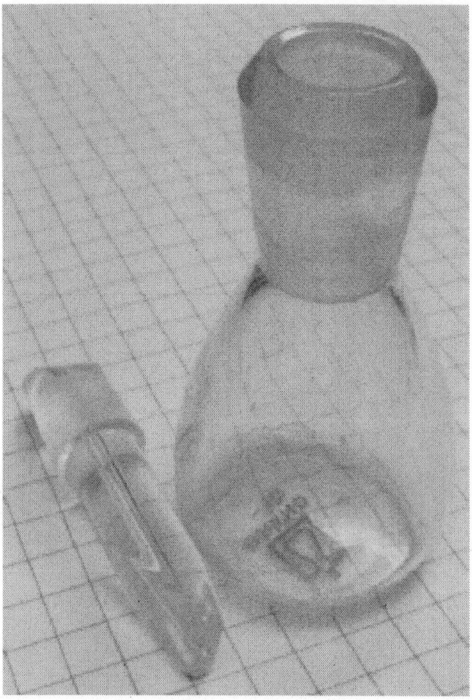

Figure: *An empty glass pycnometer and stopper*

There is also a gas-based manifestation of a pycnometre known as a *gas pycnometre.* It compares the change in pressure caused by a measured change in a closed volume containing a reference (usually a steel sphere of known volume) with the change in pressure caused by the sample under the same conditions. The difference in change of pressure represents the volume of the sample as compared to the reference sphere, and is usually used for solid particulates that may dissolve in the liquid medium of the pycnometre design described above, or for porous materials into which the liquid would not fully penetrate.

When a pycnometre is filled to a specific, but not necessarily accurately known volume, V and is placed upon a balance, it will exert a force

$$F_b = g(m_b - \rho_a \frac{m_b}{\rho_b})$$

where m_b is the mass of the bottle and g the gravitational acceleration at the location at which the measurements are being made. ρ_a is the density of the air at the ambient pressure and ρ_b is the density of the material of which the bottle is made (usually glass) so that the second term is the mass of air displaced by the glass of the bottle whose weight, by Archimedes Principle must be subtracted. The bottle is, of course, filled with air but as that air displaces an equal amount of air the weight of that air is cancelled by the weight of the air displaced. Now we fill the bottle with the reference fluid e.g. pure water. The force exerted on the pan of the balance becomes:

$$F_w = g(m_b - \rho_a \frac{m_b}{\rho_b} + V\rho_w - V\rho_a).$$

If we subtract the force measured on the empty bottle from this (or tare the balance before making the water measurement) we obtain.

$$F_{w,n} = gV(\rho_w - \rho_a)$$

where the subscript n indicated that this force is net of the force of the empty bottle. The bottle is now emptied, thoroughly dried and refilled with the sample. The force, net of the empty bottle, is now:

$$F_{s,n} = gV(\rho_s - \rho_a)$$

where ρ_s is the density of the sample. The ratio of the sample and water forces is:

$$SG_A = \frac{gV(\rho_s - \rho_a)}{gV(\rho_w - \rho_a)} = \frac{(\rho_s - \rho_a)}{(\rho_w - \rho_a)}.$$

This is called the Apparent Specific Gravity, denoted by subscript A, because it is what we would obtain if we took the ratio of net weighings in air from an analytical balance or used a hydrometre (the stem displaces air). Note that the result does not depend on the calibration of the balance. The only requirement on it is that it read linearly with force. Nor does SG_A depend on the actual volume of the pycnometre.

Further manipulation and finally substitution of SG_V, the true specific gravity (the subscript V is used because this is often referred to as the specific gravity *in vacuo*), for ρ_s/ρ_w gives the relationship between apparent and true specific gravity.

$$SG_A = \frac{\frac{\rho_s}{\rho_w} - \frac{\rho_a}{\rho_w}}{1 - \frac{\rho_a}{\rho_w}} = \frac{SG_V - \frac{\rho_a}{\rho_w}}{1 - \frac{\rho_a}{\rho_w}}$$

In the usual case we will have measured weights and want the true specific gravity. This is found from

$$SG_V = SG_A - \frac{\rho_a}{\rho_w}(SG_A - 1)$$

Since the density of dry air at 101.325 kPa at 20 °C is 0.001205 g/cm³ and that of water is 0.998203 g/cm³ we see that the difference between true and apparent specific gravities for a substance with specific gravity (20°C/20°C) of about 1.100 would be 0.000120. Where the specific gravity of the sample is close to that of water (for example dilute ethanol solutions) the correction is even smaller.

The pycnometre is used in ISO standard: ISO 1183-1:2004, ISO 1014–1985 and ASTM standard: ASTM D854.

Types

- Gay-Lussac, pear shaped, with perforated stopper, adjusted, capacity 1, 2, 5, 10, 25, 50 and 100 ml
- as above, with ground-in thermometre, adjusted, side tube with cap
- Hubbard, for bitumen and heavy oils, cylindrical type, ASTM D 70, 24 ml
- as above, conical type, ASTM D 115 and D 234, 25 ml
- Boot, with vacuum jacket and thermometre, capacity 5, 10, 25 and 50 ml

Digital Density Metres

***Hydrostatic Pressure-based Instruments*:** This technology relies upon Pascal's Principle which states that the pressure difference between two points within a vertical column of fluid is dependent upon the vertical distance between the two points, the density of the fluid and the gravitational force. This technology is often used for tank gaging applications as a convenient means of liquid level and density measure.

***Vibrating Element Transducers*:** This type of instrument requires a vibrating element to be placed in contact with the fluid of interest. The resonant frequency of the element is measured and is related to the density of the fluid by a characterization that is dependent upon the design of the element. In modern laboratories precise measurements of specific gravity are made using oscillating U-tube metres. These are capable of measurement to 5 to 6 places beyond the decimal point and are used in the brewing, distilling, pharmaceutical, petroleum and other industries. The instruments measure the actual mass of fluid contained in a fixed volume at temperatures between 0 and 80 °C but as they are microprocessor based can calculate apparent or true specific gravity and contain tables relating these to the strengths of common acids, sugar solutions, etc. The vibrating fork immersion probe is another good example of this technology. This technology also includes many coriolis-type mass flow metres which are widely used in chemical and petroleum industry for high accuracy mass flow measurement and can be configured to also output density information based on the resonant frequency of the vibrating flow tubes.

***Ultrasonic Transducer*:** Ultrasonic waves are passed from a source, through the fluid of interest, and into a detector which measures the acoustic spectroscopy of the waves. Fluid properties such as density and viscosity can be inferred from the spectrum.

***Radiation-based Gauge*:** Radiation is passed from a source, through the fluid of interest, and into a scintillation detector, or counter. As the fluid density increases, the detected radiation "counts" will decrease. The source is typically the radioactive isotope cesium-137, with a half-life of about 30 years. A key advantage for this technology is that the instrument is not required to be in contact with the fluid—typically the source and detector are mounted on the outside of tanks or piping.

***Buoyant Force Transducer*:** the buoyancy force produced by a float in a homogeneous liquid is equal to the weight of the liquid that is displaced by the float. Since buoyancy force is linear with respect to the density of the liquid within which the float is submerged, the measure of the buoyancy force yields a measure of the density of the liquid. One commercially available unit claims the instrument is capable of measuring specific gravity with an accuracy of ± 0.005 SG units. The submersible probe head contains a mathematically characterized spring-float system. When the head is immersed vertically in the liquid, the float moves vertically and the position of the float controls the position of a permanent magnet whose displacement is sensed by a concentric array of Hall-effect linear displacement sensors. The output signals of the sensors are mixed in a dedicated electronics module that provides a single output voltage whose magnitude is a direct linear measure of the quantity to be measured.

Examples

Substances with a specific gravity of 1 are neutrally buoyant, those with SG greater than one are denser than water, and so (ignoring surface tension effects) will sink in it, and those with an SG of less than one are less dense than water, and so will float.

Example:

$$RD_{H_2O} = \frac{\rho_{\text{Material}}}{\rho_{\text{H}_2\text{O}}} = SG,$$

Material	*Specific Gravity*
Balsa wood	0.2
Oak wood	0.75
Ethanol	0.78
Water	1
Table salt	2.17
Aluminium	2.7
Iron	7.87
Copper	8.96
Lead	11.35
Mercury	13.56
Depleted uranium	19.1
Gold	19.3
Osmium	22.59

(Samples may vary, and these figures are approximate.)

Viscosity

The viscosity of a fluid is a measure of its resistance to gradual deformation by shear stress or tensile stress. For liquids, it corresponds to the informal notion of "thickness". For example, honey has a higher viscosity than water.

Viscosity is due to friction between neighbouring parcels of the fluid that are moving at different velocities. When fluid is forced through a tube, the fluid generally moves faster near the axis and very slowly near the walls, therefore some stress (such as a pressure difference between the two ends of the tube) is needed to overcome the friction between layers and keep the fluid moving. For the same velocity pattern, the stress required is proportional to the fluid's viscosity. A liquid's viscosity depends on the size and shape of its particles and the attractions between the particles.

A fluid that has no resistance to shear stress is known as an ideal fluid or inviscid fluid. Zero viscosity is observed only at very low temperatures, in superfluids. Otherwise all fluids have positive viscosity. If the viscosity is very high, for instance in pitch, the fluid will appear to be a solid in the short term. A liquid whose viscosity is less than that of water is sometimes known as a mobile liquid, while a substance with a viscosity substantially greater than water is called a viscous liquid.

Etymology

The word "viscosity" is derived from the Latin "viscum", meaning mistletoe. A viscous glue called birdlime was made from mistletoe berries and was used for lime-twigs to catch birds.

Definition

Shear Viscosity:

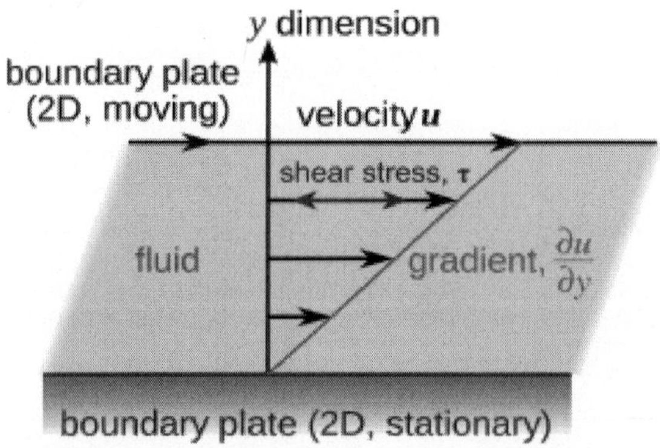

Figure: *Laminar shear of fluid between two plates. Friction between the fluid and the moving boundaries causes the fluid to shear. The force required for this action is a measure of the fluid's viscosity.*

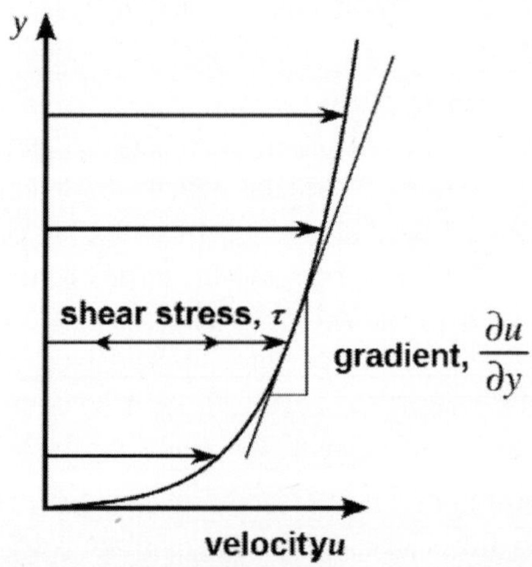

Figure: *In a general parallel flow (such as could occur in a straight pipe), the shear stress is proportional to the gradient of the velocity*

The shear viscosity of a fluid expresses its resistance to shearing flows, where adjacent layers move parallel to each other with different speeds. It can be defined through the idealized situation known as a Couette flow, where a layer of fluid is trapped between two horizontal plates, one fixed and one moving horizontally at constant speed u. (The plates are assumed to be very large, so that one need not consider what happens near their edges.)

If the speed of the top plate is small enough, the fluid particles will move parallel to it, and their speed will vary linearly from zero at the bottom to at the top. Each layer of fluid will move faster than the one just below it, and friction between them will give rise to a force resisting their relative motion. In particular, the fluid will apply on the top plate a force in the direction opposite to its motion, and an equal but opposite one to the bottom plate. An external force is therefore required in order to keep the top plate moving at constant speed.

The magnitude F of this force is found to be proportional to the speed and the area A of each plate, and inversely proportional to their separation y. That is,

$$F = \mu A \frac{u}{y}$$

The proportionality factor μ in this formula is the viscosity (specifically, the dynamic viscosity) of the fluid.

The ratio u / y is called the *rate of shear deformation* or *shear velocity*, and is the derivative of the fluid speed in the direction perpendicular to the plates. Isaac Newton expressed the viscous forces by the differential equation

$$\tau = \mu \frac{\partial u}{\partial y}$$

where $\tau = F / A$ and $\partial u / \partial y$ is the local shear velocity. This formula assumes that the flow is moving along parallel lines and the y axis, perpendicular to the flow, points in the direction of maximum shear velocity. This equation can be used where the velocity does not vary linearly with y, such as in fluid flowing through a pipe.

Use of the Greek letter mu (μ) for the dynamic stress viscosity is common among mechanical and chemical engineers, as well as physicists. However, the Greek letter eta (η) is also used by chemists, physicists, and the IUPAC.

Kinematic Viscosity

The kinematic viscosity is the dynamic viscosity μ divided by the density of the fluid ρ. It is usually denoted by the Greek letter nu (ν). It is a convenient concept when analyzing the Reynolds number, that expresses the ratio of the inertial forces to the viscous forces:

$$Re = \frac{\rho u L}{\mu} = \frac{u L}{\nu},$$

where L is a typical length scale in the system.

Bulk Viscosity

When a compressible fluid is compressed or expanded evenly, without shear, it may still exhibit a form of internal friction that resists its flow. These forces are related to the rate of compression or expansion by a factor σ, called the volume viscosity, bulk viscosity or second viscosity

The bulk viscosity is important only when the fluid is being rapidly compressed or expanded, such as in sound and shock waves. Bulk viscosity explains the loss of energy in those waves, as described by Stokes' law of sound attenuation.

Viscosity Tensor

In general, the stresses within a flow can be attributed partly to the deformation of the material from some rest state (elastic stress), and partly to the rate of change of the deformation over time (viscous stress). In a fluid, by definition, the elastic stress includes only the hydrostatic pressure.

In very general terms, the fluid's viscosity is the relation between the strain rate and the viscous stress. In the Newtonian fluid model, the relationship is by definition a linear map, described by a viscosity tensor that, multiplied by the strain rate tensor (which is the gradient of the flow's velocity), gives the viscous stress tensor.

The viscosity tensor has nine independent degrees of freedom in general. For isotropic Newtonian fluids, these can be reduced to two independent parameters. The most usual decomposition yields the stress viscosity μ and the bulk viscosity σ.

Newtonian and Non-Newtonian Fluids

Newton's law of viscosity is a constitutive equation (like Hooke's law, Fick's law, Ohm's law): it is not a fundamental law of nature but an approximation that holds in some materials and fails in others.

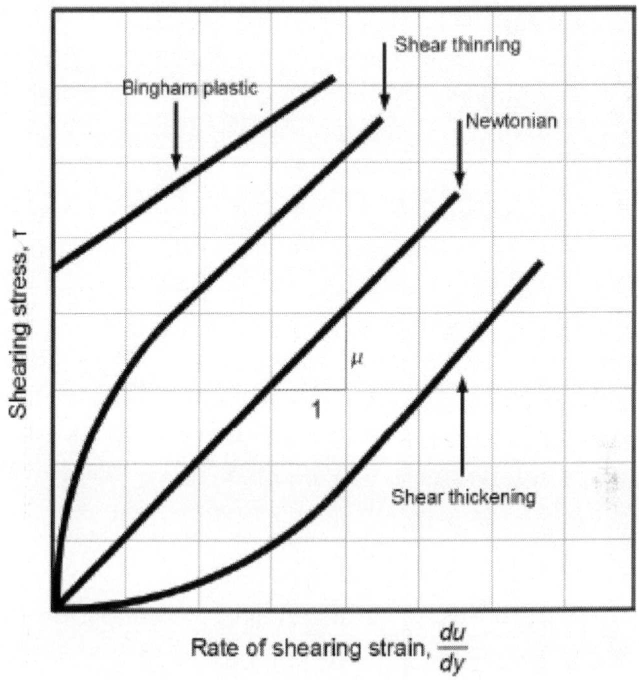

Figure: *Viscosity, the slope of each line, varies among materials*

A fluid that behaves according to Newton's law, with a viscosity μ that is independent of the stress, is said to be Newtonian. Gases, water and many common liquids can be considered Newtonian in ordinary conditions and contexts. There are many non-Newtonian fluids that significantly deviate from that law in some way or other. For example:

- Shear thickening liquids, whose viscosity increases with the rate of shear stress.
- Shear thinning liquids, whose viscosity decreases with the rate of shear stress.
- Thixotropic liquids, that become less viscous over time when shaken, agitated, or otherwise stressed.
- Rheopectic liquids, that become more viscous over time when shaken, agitated, or otherwise stressed.
- Bingham plastics that behave as a solid at low stresses but flows as a viscous fluid at high stresses.

Shear thinning liquids are very commonly, but misleadingly, described as thixotropic. Even for a Newtonian fluid, the viscosity usually depends on its composition and temperature. For gases and other compressible fluids, it depends on temperature and varies very slowly with pressure. The viscosity of some fluids may depend on other factors. A magnetorheological fluid, for example, becomes thicker when subjected to a magnetic field, possibly to the point of behaving like a solid.

Viscosity in Solids

The viscous forces that arise during fluid flow must not be confused with the elastic forces that arise in a solid in response to shear, compression or extension stresses. While in the latter the stress is proportional to the *amount* of shear deformation, in a fluid it is proportional to the *rate* of deformation over time. (For this reason, Maxwell used the term fugitive elasticity for fluid viscosity.)

However, many liquids (including water) will briefly react like elastic solids when subjected to sudden stress. Conversely, many "solids" (even granite) will flow like liquids, albeit very slowly, even under arbitrarily small stress. Such materials are therefore best described as possessing both elasticity (reaction to deformation) and viscosity (reaction to rate of deformation); that is, being viscoelastic.

Indeed, some authors have claimed that amorphous solids, such as glass and many polymers, are actually liquids with a very high viscosity (e.g.~greater than 10^{12} Pa·s). However, other authors dispute this hypothesis, claiming instead that there is some threshold for the stress, below which most solids will not flow at all, and that alleged instances of glass flow in window panes of old buildings are due to the crude manufacturing process of older eras rather than to the viscosity of glass. Viscoelastic solids may exhibit both stress viscosity and bulk viscosity. The extensional viscosity is a linear combination of the shear and bulk viscosities that describes the reaction of a solid elastic material to elongation. It is widely used for characterizing polymers. In geology, earth materials that exhibit viscous deformation at least three times greater than their elastic deformation are sometimes called rheids.

Viscosity Measurement

Viscosity is measured with various types of viscometres and rheometres. A rheometre is used for those fluids that cannot be defined by a single value of viscosity and therefore require more parameters to be set and measured than is the case for a viscometre. Close temperature control of the fluid is essential to acquire accurate measurements, particularly in materials like lubricants, whose viscosity can double with a change of only 5 °C.

For some fluids, viscosity is a constant over a wide range of shear rates (Newtonian fluids). The fluids without a constant viscosity (non-Newtonian fluids) cannot be described by a single number. Non-Newtonian fluids exhibit a variety of different correlations between shear stress and shear rate.

One of the most common instruments for measuring kinematic viscosity is the glass capillary viscometre.

In paint industries, viscosity is commonly measured with a Zahn cup, in which the efflux time is determined and given to customers. The efflux time can also be converted to kinematic viscosities (centistokes, cSt) through the conversion equations.

Also used in paint, a Stormer viscometre uses load-based rotation in order to determine viscosity. The viscosity is reported in Krebs units (KU), which are unique to Stormer viscometres.

A Ford viscosity cup measures the rate of flow of a liquid. This, under ideal conditions, is proportional to the kinematic viscosity.

Vibrating viscometres can also be used to measure viscosity. These models such as the *Dynatrol* use vibration rather than rotation to measure viscosity.

Extensional viscosity can be measured with various rheometres that apply extensional stress.

Volume viscosity can be measured with an acoustic rheometre.

Apparent viscosity is a calculation derived from tests performed on drilling fluid used in oil or gas well development. These calculations and tests help engineers develop and maintain the properties of the drilling fluid to the specifications required.

Units

Dynamic Viscosity: The SI physical unit of dynamic viscosity is the pascal-second (Pa·s), (equivalent to N·s/m^2, or kg/(m·s)). If a fluid with a viscosity of one Pa·s is placed between two plates, and one plate is pushed sideways with a shear stress of one pascal, it moves a distance equal to the thickness of the layer between the plates in one second. Water at 20 °C has a viscosity of 0.001002 Pa·s, while a typical motor oil could have a viscosity of about 0.250 Pa·s.

The cgs physical unit for dynamic viscosity is the *poise* (P), named after Jean Louis Marie Poiseuille. It is more commonly expressed, particularly in ASTM standards, as *centipoise* (cP). Water at 20 °C has a viscosity of 1.0020 cP.

$1\ \text{P} = 0.1\ \text{Pa·s},$

$1\ \text{cP} = 1\ \text{mPa·s} = 0.001\ \text{Pa·s}.$

Kinematic Viscosity

The SI unit of kinematic viscosity is m^2/s.

The cgs physical unit for kinematic viscosity is the *stokes* (St), named after George Gabriel Stokes. It is sometimes expressed in terms of *centistokes* (cSt). In U.S. usage, *stoke* is sometimes used as the singular form.

$1\ \text{St} = 1\ \text{cm}^2 \cdot \text{s}^{-1} = 10^{-4}\ \text{m}^2 \cdot \text{s}^{-1}.$

$1\ \text{cSt} = 1\ \text{mm}^2 \cdot \text{s}^{-1} = 10^{-6} \text{m}^2 \cdot \text{s}^{-1}.$

Water at 20 °C has a kinematic viscosity of about 1 cSt. The kinematic viscosity is sometimes referred to as diffusivity of momentum, because it is analogous to diffusivity of heat and diffusivity of mass. It is therefore used in dimensionless numbers which compare the ratio of the diffusivities.

Fluidity

The reciprocal of viscosity is *fluidity*, usually symbolized by $\varphi = 1 / \mu$ or $F = 1 / \mu$, depending on the convention used, measured in *reciprocal poise* (cm·s·g^{-1}), sometimes called the *rhe*. *Fluidity* is seldom used in engineering practice. The concept of fluidity can be used to determine the viscosity of an ideal solution. For two components *a* and *b*, the fluidity when *a* and *b* are mixed is

$$F \approx \chi_a F_a + \chi_b F_b, ,$$

which is only slightly simpler than the equivalent equation in terms of viscosity:

$$\mu \approx \frac{1}{\chi_a / \mu_a + \chi_b / \mu_b},$$

where χ_a and χ_b is the mole fraction of component *a* and *b* respectively, and μ_a and μ_b are the components' pure viscosities.

Non-Standard Units

The Reyn is a British unit of dynamic viscosity. Viscosity index is a measure for the change of kinematic viscosity with temperature. It is used to characterise lubricating oil in the automotive industry.

At one time the petroleum industry relied on measuring kinematic viscosity by means of the Saybolt viscometre, and expressing kinematic viscosity in units of *Saybolt Universal Seconds* (SUS). Other abbreviations such as SSU (*Saybolt Seconds Universal*) or SUV (*Saybolt Universal Viscosity*) are sometimes used. Kinematic viscosity in centistoke can be converted from SUS according to the arithmetic and the reference table provided in ASTM D 2161.

Molecular Origins

The viscosity of a system is determined by how molecules constituting the system interact. There are no simple but correct expressions for the viscosity of a fluid. The simplest exact expressions are the Green–Kubo relations for the linear shear viscosity or the Transient Time Correlation Function expressions derived by Evans and Morriss in 1985. Although these expressions are each exact, in order to calculate the viscosity of a dense fluid using these relations currently requires the use of molecular dynamics computer simulations.

Gases

Viscosity in gases arises principally from the molecular diffusion that transports momentum between layers of flow. The kinetic theory of gases allows accurate prediction of the behaviour of gaseous viscosity.

Within the regime where the theory is applicable:

- Viscosity is independent of pressure and
- Viscosity increases as temperature increases.

James Clerk Maxwell published a famous paper in 1866 using the kinetic theory of gases to study gaseous viscosity. To understand why the viscosity is independent of pressure, consider two adjacent boundary layers (A and B) moving with respect to each other. The internal friction (the viscosity) of the gas is determined by the probability a particle of layer A enters layer B with a corresponding transfer of momentum. Maxwell's calculations show that the viscosity coefficient is proportional to the density, the mean free path, and the mean velocity of the atoms. On the other hand, the *mean free path* is inversely proportional to the density. So an increase in density due to an increase in pressure doesn't result in any change in viscosity.

Relation to Mean Free Path of Diffusing Particles

In relation to diffusion, the kinematic viscosity provides a better understanding of the behaviour of mass transport of a dilute species. Viscosity is related to shear stress and the rate of shear in a fluid, which illustrates its dependence on the mean free path, λ, of the diffusing particles.

From fluid mechanics, for a Newtonian fluid, the shear stress, τ, on a unit area moving parallel to itself, is found to be proportional to the rate of change of velocity with distance perpendicular to the unit area:

$$\tau = \mu \frac{du_x}{dy}$$

for a unit area parallel to the x-z plane, moving along the x axis. We will derive this formula and show how μ is related to λ. Interpreting shear stress as the time rate of change of momentum, p, per unit area A (rate of momentum flux) of an arbitrary control surface gives

$$\tau = \frac{\dot{p}}{A} = \frac{\dot{m}\langle u_x \rangle}{A}.$$

where $\langle u_x \rangle$ is the average velocity, along the x axis, of fluid molecules hitting the unit area, with respect to the unit area.

Further manipulation will show

$$\dot{m} = \rho \bar{u} A$$

$\langle u_x \rangle = \frac{1}{2}\lambda \frac{du_x}{dy}$, assuming that molecules hitting the unit area come from all distances between 0 and λ (equally distributed), and that their average velocities change linearly with distance (always true for small enough λ). From this follows:

$$\tau = \underbrace{\frac{1}{2}\rho \bar{u} \lambda}_{\mu} \cdot \frac{du_x}{dy} \Rightarrow \nu = \frac{\mu}{\rho} = \tfrac{1}{2}\bar{u}\lambda,$$

where

$\dot{m}$ is the rate of fluid mass hitting the surface,

ρ is the density of the fluid,

$\hat{u}$ is the average molecular speed ($\bar{u} = \sqrt{\langle u^2 \rangle}$),

μ is the dynamic viscosity.

Effect of Temperature on the Viscosity of a Gas

Sutherland's formula *can be used to derive the dynamic viscosity of an ideal gas as a function of the temperature:*

$$\mu = \mu_0 \frac{T_0 + C}{T + C} \left(\frac{T}{T_0} \right)^{3/2}.$$

This in turn is equal to

$\lambda \frac{T^{3/2}}{T+C}$, where $\lambda = \frac{\mu_0 (T_0 + C)}{T_0^{3/2}}$ is a constant for the gas.

in Sutherland's formula:

- μ = dynamic viscosity in (Pa·s) at input temperature T,
- μ_0 = reference viscosity in (Pa·s) at reference temperature T_0,
- T = input temperature in kelvins,
- T_0 = reference temperature in kelvins,
- C = Sutherland's constant for the gaseous material in question.

Valid for temperatures between $0 < T < 555$ K with an error due to pressure less than 10% below 3.45 MPa.

According to Sutherland's formula, if the absolute temperature is less than C, the relative change in viscosity for a small change in temperature is greater than the relative change in the absolute temperature, but it is smaller when T is above C. The kinematic viscosity though always increases faster than the temperature (that is, d log(ν)/d log(T) is greater than 1).

Sutherland's constant, reference values and λ values for some gases:

Gas	***C [K]***	***T₀ [K]***	***μ₀ [μPa s]***	***λ [μPa s K$^{-1/2}$]***
air	120	291.15	18.27	1.512041288
nitrogen	111	300.55	17.81	1.406732195
oxygen	127	292.25	20.18	1.693411300
carbon dioxide	240	293.15	14.8	1.572085931
carbon monoxide	118	288.15	17.2	1.428193225
hydrogen	72	293.85	8.76	0.636236562
ammonia	370	293.15	9.82	1.297443379
sulfur dioxide	416	293.65	12.54	1.768466086
helium	79.4	273	19	1.484381490

Viscosity of a Dilute Gas

The Chapman-Enskog equation may be used to estimate viscosity for a dilute gas. This equation is based on a semi-theoretical assumption by Chapman and Enskog. The equation requires three empirically determined parameters: the collision diameter (σ), the maximum energy of attraction divided by the Boltzmann constant (ϵ/κ) and the collision integral ($\omega(T^*)$).

$$\mu_0 \times 10^6 = 2.6693 \frac{(MT)^{1/2}}{\sigma^2 \omega(T^*)},$$

with

- $T^* = \kappa T/\varepsilon$ — reduced temperature (dimensionless),
- μ_0 = viscosity for dilute gas (μPa.s),
- M = molecular mass (g/mol),
- T = temperature (K),
- σ = the collision diameter (Å),
- ε / κ = the maximum energy of attraction divided by the Boltzmann constant (K),
- ω_μ = the collision integral.

Liquids

In liquids, the additional forces between molecules become important. This leads to an additional contribution to the shear stress though the exact mechanics of this are still controversial. Thus, in liquids:

- Viscosity is independent of pressure (except at very high pressure); and
- Viscosity tends to fall as temperature increases (for example, water viscosity goes from 1.79 cP to 0.28 cP in the temperature range from 0 °C to 100 °C).

The dynamic viscosities of liquids are typically several orders of magnitude higher than dynamic viscosities of gases.

Viscosity of Blends of Liquids

The viscosity of the blend of two or more liquids can be estimated using the Refutas equation. The calculation is carried out in three steps.

The first step is to calculate the Viscosity Blending Number (VBN) (also called the Viscosity Blending Index) of each component of the blend:

(1) $\text{VBN} = 14.534 \times \ln[\ln(\nu + 0.8)] + 10.975$

where ν is the kinematic viscosity in centistokes (cSt). It is important that the kinematic viscosity of each component of the blend be obtained at the same temperature.

The next step is to calculate the VBN of the blend, using this equation:

(2) $\text{VBN}_{\text{Blend}} = [x_A \times \text{VBN}_A] + [x_B \times \text{VBN}_B] + \cdots + [x_N \times \text{VBN}_N]$ where x_X is the mass fraction of each component of the blend.

Once the viscosity blending number of a blend has been calculated using equation (2), the final step is to determine the kinematic viscosity of the blend by solving equation (1) for ν:

(3) $\nu = \exp\left(\exp\left(\dfrac{\text{VBN}_{\text{Blend}} - 10.975}{14.534}\right)\right) - 0.8$, where VBN_{Blend} is the viscosity blending number of the blend.

Fluid Statistics

Pascal's Law

Pascal's law or the principle of transmission of fluid-pressure is a principle in fluid mechanics that states that pressure exerted anywhere in a confined incompressible fluid is transmitted equally in all directions throughout the fluid such that the pressure ratio (initial difference) remains the same. The law was established by French mathematician Blaise Pascal.

Definition

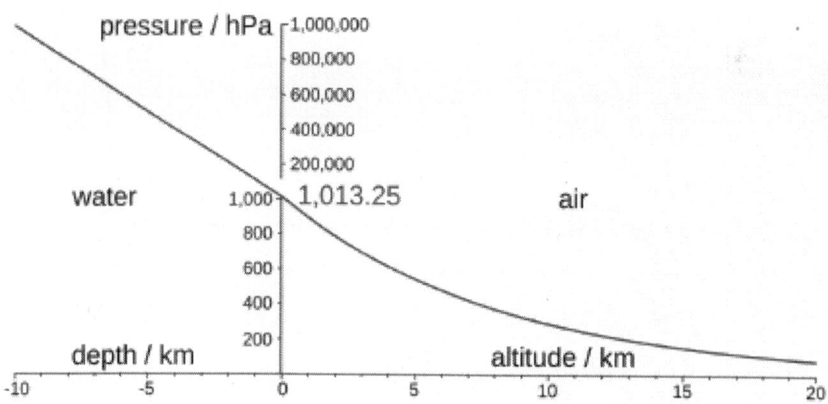

Figure: *Pressure in water and air. Pascal's law applies only for fluids.*

A change in pressure at any point in an enclosed fluid at rest is transmitted undiminished to all points in the fluid

This principle is stated mathematically as:

$$\Delta P = \rho g(\Delta h)$$

ΔP is the hydrostatic pressure (given in pascals in the SI system), or the difference in pressure at two points within a fluid column, due to the weight of the fluid;

ρ is the fluid density (in kilograms per cubic metre in the SI system);

g is acceleration due to gravity (normally using the sea level acceleration due to Earth's gravity in metres per second squared);

Δh is the height of fluid above the point of measurement, or the difference in elevation between the two points within the fluid column (in metres in SI).

The intuitive explanation of this formula is that the change in pressure between two elevations is due to the weight of the fluid between the elevations. Note that the variation with height does not depend on any additional pressures. Therefore Pascal's law can be interpreted as saying that *any change in pressure applied* at any given point of the fluid is transmitted *undiminished throughout* the fluid.

Explanation

If a U-tube is filled with water and pistons are placed at each end, pressure exerted against the left piston will be transmitted throughout the liquid and against the bottom of the right piston. (The pistons are simply "plugs" that can slide freely but snugly inside the tube.) The pressure that the left piston exerts against the water will be exactly equal to the pressure the water exerts against the right piston. Suppose the tube on the right side is made wider and a piston of a larger area is used; for example, the piston on the right has 50 times the area of the piston on the left. If a 1 N load is placed on the left piston, an additional pressure (nearly 1 N/cm^2) due to the weight of the load is transmitted throughout the liquid and up against the larger piston. The difference between force and pressure is important: the additional pressure is exerted against every square centimetre of the larger piston. Since there is 50 times the area, 50 times as much force is exerted on the larger piston. Thus, the larger piston will support a 50 N load - fifty times the load on the smaller piston.

Forces can be multiplied using such a device. One newton input produces 50 newtons output. By further increasing the area of the larger piston (or reducing the area of the smaller piston), forces can be multiplied, in principle, by any amount. Pascal's principle underlies the operation of the hydraulic press. The hydraulic press does not violate energy conservation, because a decrease in distance moved compensates for the increase in force. When the small piston is moved downward 10 centimetres, the large piston will be raised only one-fiftieth of this, or 0.2 centimetres. The input force multiplied by the distance moved by the smaller piston is equal to the output force multiplied by the distance moved by the larger piston; this is one more example of a simple machine operating on the same principle as a mechanical lever.

Pascal's principle applies to all fluids, whether gases or liquids. A typical application of Pascal's principle for gases and liquids is the automobile lift seen in many service stations (the hydraulic jack). Increased air pressure produced by an air compressor is transmitted through the air to the surface of oil in an underground reservoir. The oil, in turn, transmits the pressure to a piston, which lifts the automobile. The relatively low pressure that exerts the lifting force against the piston is about the same as the air pressure in automobile tires. Hydraulics is employed by modern devices ranging from very small to enormous. For example, there are hydraulic pistons in almost all construction machines where heavy loads are involved.

Applications

- The siphon
- The underlying principle of the hydraulic jack and hydraulic press
- Force amplification in the braking system of most motor vehicles.
- Used in artesian wells, water towers, and dams.
- Scuba divers must understand this principle. At a depth of 10 metres under water, pressure is twice the atmospheric pressure at sea level, and increases by about 100 kPa for each increase of 10 m depth.

Pressure Measurement

Many techniques have been developed for the measurement of pressure and vacuum. Instruments used to measure pressure are called pressure gauges or vacuum gauges.

A manometre could also refer to a pressure measuring instrument, usually limited to measuring pressures near to atmospheric. The term *manometre* is often used to refer specifically to liquid column hydrostatic instruments. A vacuum gauge is used to measure the pressure in a vacuum—which is further divided into two subcategories, high and low vacuum (and sometimes ultra-high vacuum). The applicable pressure range of many of the techniques used to measure vacuums have an overlap. Hence, by combining several different types of gauge, it is possible to measure system pressure continuously from 10 mbar down to 10^{-11} mbar.

Absolute, Gauge and Differential Pressures - Zero Reference

Everyday pressure measurements, such as for tire pressure, are usually made relative to ambient air pressure. In other cases measurements are made relative to a vacuum or to some other specific reference. When distinguishing between these zero references, the following terms are used:

- Absolute pressure is zero-referenced against a perfect vacuum, so it is equal to gauge pressure plus atmospheric pressure.
- Gauge pressure is zero-referenced against ambient air pressure, so it is equal to absolute pressure minus atmospheric pressure. Negative signs are usually omitted. To distinguish a negative pressure, the value may be appended with the word "vacuum" or the gauge may be labelled a "vacuum gauge."
- Differential pressure is the difference in pressure between two points.

The zero reference in use is usually implied by context, and these words are added only when clarification is needed. Tire pressure and blood pressure are gauge pressures by convention, while atmospheric pressures, deep vacuum pressures, and altimetre pressures must be absolute.

For most working fluids where a fluid exists in a closed system, gauge pressure measurement prevails. Pressure instruments connected to the system will indicate pressures relative to the current atmospheric pressure. The situation changes when extreme vacuum pressures are measured; absolute pressures are typically used instead.

Differential pressures are commonly used in industrial process systems. Differential pressure gauges have two inlet ports, each connected to one of the volumes whose pressure is to be monitored. In effect, such a gauge performs the mathematical operation of subtraction through mechanical means, obviating the need for an operator or control system to watch two separate gauges and determine the difference in readings. Moderate vacuum pressure readings can be ambiguous without the proper context, as they may represent absolute pressure or gauge pressure without a negative sign. Thus a vacuum of 26 inHg gauge is equivalent to an absolute pressure of 30 inHg (typical atmospheric pressure) – 26 inHg = 4 inHg.

Atmospheric pressure is typically about 100 kPa at sea level, but is variable with altitude and weather. If the absolute pressure of a fluid stays constant, the gauge pressure of the same fluid will vary as atmospheric pressure changes. For example, when a car drives up a mountain, the (gauge) tire pressure goes up because atmospheric pressure goes down. The absolute pressure in the tire is essentially unchanged.

Using atmospheric pressure as reference is usually signified by a *g* for gauge after the pressure unit, e.g. 70 psig, which means that the pressure measured is the total pressure minus atmospheric pressure. There are two types of gauge reference pressure: vented gauge (vg) and sealed gauge (sg).

A vented gauge pressure transmitter for example allows the outside air pressure to be exposed to the negative side of the pressure sensing diaphragm, via a vented cable or a hole on the side of the device, so that it always measures the pressure referred to ambient barometric pressure. Thus a vented gauge reference pressure sensor should always read zero pressure when the process pressure connection is held open to the air.

A sealed gauge reference is very similar except that atmospheric pressure is sealed on the negative side of the diaphragm. This is usually adopted on high pressure ranges such as hydraulics where atmospheric pressure changes will have a negligible effect on the accuracy of the reading, so venting is not necessary. This also allows some manufacturers to provide secondary pressure containment as an extra precaution for pressure equipment safety if the burst pressure of the primary pressure sensing diaphragm is exceeded.

There is another way of creating a sealed gauge reference and this is to seal a high vacuum on the reverse side of the sensing diaphragm. Then the output signal is offset so the pressure sensor reads close to zero when measuring atmospheric pressure.

A sealed gauge reference pressure transducer will never read exactly zero because atmospheric pressure is always changing and the reference in this case is fixed at 1 bar.

An absolute pressure measurement is one that is referred to absolute vacuum. The best example of an absolute referenced pressure is atmospheric or barometric pressure.

To produce an absolute pressure sensor the manufacturer will seal a high vacuum behind the sensing diaphragm. If the process pressure connection of an absolute pressure transmitter is open to the air, it will read the actual barometric pressure.

Units

Pressure units						
• v • t • e	*Pascal* *(Pa)*	*Bar* *(bar)*	*Technical atmosphere* *(at)*	*Standard atmosphere* *(atm)*	*Torr* *(Torr)*	*Pounds per square inch* *(psi)*
1 Pa	≡ 1 N/m^2	10^{-5}	1.0197×10^{-5}	9.8692×10^{-6}	7.5006×10^{-3}	1.450377×10^{-4}
1 bar	10^5	≡ 10^6 dyn/cm^2	1.0197	0.98692	750.06	14.50377
1 at	0.980665×10^5	0.980665	≡ 1 kp/cm^2	0.9678411	735.5592	14.22334
1 atm	1.01325×10^5	1.01325	1.0332	≡ p_0	≡ 760	14.69595
1 Torr	133.3224	1.333224×10^{-3}	1.359551×10^{-3}	1.315789×10^{-3}	≈ 1 mm$_{Hg}$	1.933678×10^{-2}
1 psi	6.8948×10^3	6.8948×10^{-2}	7.03069×10^{-2}	6.8046×10^{-2}	51.71493	≡ 1 lb$_F$/in^2

The SI unit for pressure is the pascal (Pa), equal to one newton per square metre (N·m^{-2} or kg·m^{-1}·s^{-2}). This special name for the unit was added in 1971; before that, pressure in SI was expressed in units such as N·m^{-2}. When indicated, the zero reference is stated in parenthesis following the unit, for example 101 kPa (abs). The pound per square inch (psi) is still in widespread use in the US and Canada, for measuring, for instance, tire pressure. A letter is often appended to the psi unit to indicate the measurement's zero reference; psia for absolute, psig for gauge, psid for differential, although this practice is discouraged by the NIST.

Because pressure was once commonly measured by its ability to displace a column of liquid in a manometre, pressures are often expressed as a depth of a particular fluid (*e.g.*, inches of water). Manometric measurement is the subject of pressure head calculations. The most common choices for a manometre's fluid are mercury (Hg) and water; water is nontoxic and readily available, while mercury's density allows for a shorter column (and so a smaller manometre) to measure a given pressure. The abbreviation "W.C." or the words "water column" are often printed on gauges and measurements that use water for the manometre. Fluid density and local gravity can vary from one reading to another depending on local factors, so the height of a fluid column does not define pressure precisely. So measurements in "millimetres of mercury" or "inches of mercury" can be converted to SI units as long as attention is paid to the local factors of fluid density and gravity. Temperature fluctuations change the value of fluid density, while location can affect gravity.

Although no longer preferred, these manometric units are still encountered in many fields. Blood pressure is measured in millimetres of mercury in most of the world, and lung pressures in centimetres of water are still common, as in settings for CPAP machines. Natural gas pipeline pressures are measured in inches of water, expressed as "inches W.C." Scuba divers often use a manometric rule of thumb: the pressure exerted by ten metres depth of water is approximately equal to one atmosphere. In vacuum systems, the units torr, micrometre of mercury (micron), and inch of mercury (inHg) are most commonly used. Torr and micron usually indicates an absolute pressure, while inHg usually indicates a gauge pressure.

Atmospheric pressures are usually stated using kilopascal (kPa), or atmospheres (atm), except in American meteorology where the hectopascal (hPa) and millibar (mbar) are preferred. In American and Canadian engineering, stress is often measured in kip. Note that stress is not a true pressure since it is not scalar. In the cgs system the unit of pressure was the barye (ba), equal to 1 dyn·cm^{-2}. In the mts system, the unit of pressure was the pieze, equal to 1 sthene per square metre. Many other hybrid units are used such as mmHg/cm^2 or grams-force/cm^2 (sometimes as [[kg/cm^2]] without properly identifying the force units). Using the names kilogram, gram, kilogram-force, or gram-force (or their symbols) as a unit of force is prohibited in SI; the unit of force in SI is the newton (N).

Static and Dynamic Pressure

Static pressure is uniform in all directions, so pressure measurements are independent of direction in an immovable (static) fluid. Flow, however, applies additional pressure on surfaces perpendicular to the flow direction, while having little impact on surfaces parallel to the flow direction. This directional component of pressure in a moving (dynamic) fluid is called dynamic pressure. An instrument facing the flow direction measures the sum of the static and dynamic pressures; this measurement is called the total pressure or stagnation pressure. Since dynamic pressure is referenced to static pressure, it is neither gauge nor absolute; it is a differential pressure. While static gauge pressure is of primary importance to determining net loads on pipe walls, dynamic pressure is used to measure flow rates and airspeed. Dynamic pressure can be measured by taking the differential pressure between instruments parallel and perpendicular to the flow. Pitot-static tubes, for example perform this measurement on airplanes to determine airspeed. The presence of the measuring instrument inevitably acts to divert flow and create turbulence, so its shape is critical to accuracy and the calibration curves are often non-linear.

Applications

- Altimetre
- Barometre
- MAP sensor
- Pitot tube
- Sphygmomanometre

Instruments

Many instruments have been invented to measure pressure, with different advantages and disadvantages. Pressure range, sensitivity, dynamic response and cost all vary by several orders of magnitude from one instrument design to the next. The oldest type is the liquid column (a vertical tube filled with mercury) manometre invented by Evangelista Torricelli in 1643. The U-Tube was invented by Christian Huygens in 1661.

Hydrostatic

Hydrostatic gauges (such as the mercury column manometre) compare pressure to the hydrostatic force per unit area at the base of a column of fluid. Hydrostatic gauge measurements are independent of the type of gas being measured, and can be designed to have a very linear calibration. They have poor dynamic response.

Piston

Piston-type gauges counterbalance the pressure of a fluid with a spring (for example tire-pressure gauges of comparatively low accuracy) or a solid weight, in which case it is known as a deadweight tester and may be used for calibration of other gauges.

Liquid Column

By using Bernoulli's principle and the derived pressure head equation, liquids can be used for instrumentation where gravity is present. Liquid column gauges consist of a vertical column of liquid in a tube that has ends which are exposed to different pressures. The column will rise or fall until its weight (a force applied due to gravity) is in equilibrium with the pressure differential between the two ends of the tube (a force applied due to fluid pressure). A very simple version is a U-shaped tube half-full of liquid, one side of which is connected to the region of interest while the reference pressure (which might be the atmospheric pressure or a vacuum) is applied to the other. The difference in liquid level represents the applied pressure. The pressure exerted by a column of fluid of height h and density ρ is given by the hydrostatic pressure equation, $P = hg\rho$. Therefore the pressure difference between the applied pressure P_a and the reference pressure P_0 in a U-tube manometre can be found by solving $P_a - P_0 = hg\rho$. In other words, the pressure on either end of the liquid (shown in blue in the figure to the right) must be balanced (since the liquid is static) and so $P_a = P_0 + hg\rho$.

In most liquid column measurements, the result of the measurement is the height, h, expressed typically in mm, cm, or inches. The h is also known as the pressure head. When expressed as a pressure head, pressure is specified in units of length and the measurement fluid must be specified. When accuracy is critical, the temperature of the measurement fluid must likewise be specified, because liquid density is a function of temperature. So, for example, pressure head might be written "742.2 mm_{Hg}" or "4.2 in_{H2O} at 59 °F" for measurements taken with mercury or water as the manometric fluid, respectively. The word "gauge" or "vacuum" may be added to such a measurement to distinguish between a pressure above or below the atmospheric pressure. Both mm of mercury and inches of water are common pressure heads which can be converted to S.I. units of pressure using unit conversion and the above formulas.

If the fluid being measured is significantly dense, hydrostatic corrections may have to be made for the height between the moving surface of the manometre working fluid and the location where the pressure measurement is desired except when measuring differential pressure of a fluid (for example across an orifice plate or venturi), in which case the density ρ should be corrected by subtracting the density of the fluid being measured.

To measure the pressure of a fluid accurately using a liquid column, the fluid being measured should not be flowing for a static pressure measurement. A column connected to a flowing fluid will measure static plus dynamic pressure. So if a fluid is flowing, the liquid column will change due to dynamic pressure, proportional to the square of the fluid's velocity. This of course is precisely the desired measurement when a differential pressure measurement is needed for a venturi or an orifice plate. Measuring dynamic pressures is commonly used as an intermediary in determining a fluid's velocity or flow rate.

As an example, an airplane flying through the air at sea level would experience the atmospheric pressure as a static pressure exerted on the skin of the aircraft. However, the forward surfaces of an aircraft in flight experience dynamic pressure in addition to the static pressure. To measure the static air pressure, we use a barometre in still air. To measure the dynamic pressure, imagine a mercury manometre like the U-tube above with an open end pointing in the direction of the airplane's travel and a closed end kept at the static air pressure. Mercury is pushed down the tube farther than it would if only measuring still air. For a plane travelling around 129 m/s, the dynamic pressure adds about 10% to the atmospheric pressure at sea level. A U-tube for measuring dynamic pressure on an airplane would be impractical, so a pitot tube is used instead that relies on a diaphragm rather than columns of fluid. Although dynamic pressure can be measured directly, fluid speed and air speed can be measured indirectly using the Bernoulli principle if both dynamic and static pressures are known.

Although any fluid can be used, mercury is preferred for its high density (13.534 g/cm^3) and low vapour pressure. For low pressure differences well above the vapour pressure of water, water is commonly used (and "inches of waterGgauge

A McLeod gauge isolates a sample of gas and compresses it in a modified mercury manometre until the pressure is a few mmHg. The gas must be well-behaved during its compression (it must not condense, for example). The technique is slow and unsuited to continual monitoring, but is capable of good accuracy.

Useful range: above 10^{-4} torr (roughly 10^{-2} Pa) as high as 10^{-6} Torr (0.1 mPa),

0.1 mPa is the lowest direct measurement of pressure that is possible with current technology. Other vacuum gauges can measure lower pressures, but only indirectly by measurement of other pressure-controlled properties. These indirect measurements must be calibrated to SI units via a direct measurement, most commonly a McLeod gauge.

Aneroid

Aneroid gauges are based on a metallic pressure sensing element that flexes elastically under the effect of a pressure difference across the element. "Aneroid" means "without fluid," and the term originally distinguished these gauges from the hydrostatic gauges described above. However, aneroid gauges can be used to measure the pressure of a liquid as well as a gas, and they are not the only type of gauge that can operate without fluid. For this reason, they are often called mechanical gauges in modern language. Aneroid gauges are not dependent on the type of gas being measured, unlike thermal and ionization gauges, and are less likely to contaminate the system than hydrostatic gauges. The pressure sensing element may be a Bourdon tube, a diaphragm, a capsule, or a set of bellows, which will change shape in response to the pressure of the region in question. The deflection of the pressure sensing element may be read by a linkage connected to a needle, or it may be read by a secondary transducer. The most common secondary transducers in modern vacuum gauges measure a change in capacitance due to the mechanical deflection. Gauges that rely on a change in capacitances are often referred to as Baratron gauges.

Bourdon

The Bourdon pressure gauge uses the principle that a flattened tube tends to straighten or regain its circular form in cross-section when pressurized. Although this change in cross-section may be hardly noticeable, and thus involving moderate stresses within the elastic range of easily workable materials, the strain of the material of the tube is magnified by forming the tube into a C shape or even a helix, such that the entire tube tends to straighten out or uncoil, elastically, as it is pressurized. Eugene Bourdon patented his gauge in France in 1849, and it was widely adopted because of its superior sensitivity, linearity, and accuracy; Edward Ashcroft purchased Bourdon's American patent rights in 1852 and became a major manufacturer of gauges. Also in 1849, Bernard Schaeffer in Magdeburg, Germany patented a successful diaphragm pressure gauge, which, together with the Bourdon gauge, revolutionized pressure measurement in industry. But in 1875 after Bourdon's patents expired, his company Schaeffer and Budenberg also manufactured Bourdon tube gauges.

In practice, a flattened thin-wall, closed-end tube is connected at the hollow end to a fixed pipe containing the fluid pressure to be measured. As the pressure increases, the closed end moves in an arc, and this motion is converted into the rotation of a (segment of a) gear by a connecting link that is usually adjustable. A small-diameter pinion gear is on the pointer shaft, so the motion is magnified further by the gear ratio. The positioning of the indicator card behind the pointer, the initial pointer shaft position, the linkage length and initial position, all provide means to calibrate the pointer to indicate the desired range of pressure for variations in the behaviour of the Bourdon tube itself. Differential pressure can be measured by gauges containing two different Bourdon tubes, with connecting linkages.

Bourdon tubes measure gauge pressure, relative to ambient atmospheric pressure, as opposed to absolute pressure; vacuum is sensed as a reverse motion. Some aneroid barometres use Bourdon tubes closed at both ends (but most use diaphragms or capsules). When the measured pressure is rapidly pulsing, such as when the gauge is near a reciprocating pump, an orifice restriction in the connecting pipe is frequently used to avoid unnecessary wear on the gears and provide an average reading; when the whole gauge is subject to mechanical vibration, the entire case including the pointer and indicator card can be filled with an oil or glycerin. Tapping on the face of the gauge is not recommended as it will tend to falsify actual readings initially presented by the gauge. The Bourdon tube is separate from the face of the gauge and thus has no effect on the actual reading of pressure. Typical high-quality modern gauges provide an accuracy of ±2% of span, and a special high-precision gauge can be as accurate as 0.1% of full scale.

In the following illustrations the transparent cover face of the pictured combination pressure and vacuum gauge has been removed and the mechanism removed from the case. This particular gauge is a combination vacuum and pressure gauge used for automotive diagnosis:

- the left side of the face, used for measuring manifold vacuum, is calibrated in centimetres of mercury on its inner scale and inches of mercury on its outer scale.
- the right portion of the face is used to measure fuel pump pressure and is calibrated in fractions of 1 kgf/cm^2 on its inner scale and pounds per square inch on its outer scale.

Mechanical Details

Stationary parts:

- A: Receiver block. This joins the inlet pipe to the fixed end of the Bourdon tube (1) and secures the chassis plate (B). The two holes receive screws that secure the case.
- B: Chassis plate. The face card is attached to this. It contains bearing holes for the axles.
- C: Secondary chassis plate. It supports the outer ends of the axles.
- D: Posts to join and space the two chassis plates.

Moving Parts:

1. Stationary end of Bourdon tube. This communicates with the inlet pipe through the receiver block.

2. Moving end of Bourdon tube. This end is sealed.
3. Pivot and pivot pin.
4. Link joining pivot pin to lever (5) with pins to allow joint rotation.
5. Lever. This is an extension of the sector gear (7).
6. Sector gear axle pin.
7. Sector gear.
8. Indicator needle axle. This has a spur gear that engages the sector gear (7) and extends through the face to drive the indicator needle. Due to the short distance between the lever arm link boss and the pivot pin and the difference between the effective radius of the sector gear and that of the spur gear, any motion of the Bourdon tube is greatly amplified. A small motion of the tube results in a large motion of the indicator needle.
9. Hair spring to preload the gear train to eliminate gear lash and hysteresis.

Diaphragm

A second type of aneroid gauge uses the deflection of a flexible membrane that separates regions of different pressure. The amount of deflection is repeatable for known pressures so the pressure can be determined by using calibration. The deformation of a thin diaphragm is dependent on the difference in pressure between its two faces. The reference face can be open to atmosphere to measure gauge pressure, open to a second port to measure differential pressure, or can be sealed against a vacuum or other fixed reference pressure to measure absolute pressure. The deformation can be measured using mechanical, optical or capacitive techniques. Ceramic and metallic diaphragms are used.

Useful range: above 10^{-2} Torr (roughly 1 Pa)

For absolute measurements, welded pressure capsules with diaphragms on either side are often used.

shape:

- Flat
- corrugated
- flattened tube
- capsule

Bellows

In gauges intended to sense small pressures or pressure differences, or require that an absolute pressure be measured, the gear train and needle may be driven by an enclosed and sealed bellows chamber, called an aneroid, which means "without liquid". (Early barometres used a column of liquid such as water or the liquid metal mercury suspended by a vacuum.) This bellows configuration is used in aneroid barometres (barometres with an indicating needle and dial card), altimetres, altitude recording barographs, and the altitude telemetry instruments used in weather balloon radiosondes. These devices use the sealed chamber as a reference pressure and are driven by the external pressure. Other sensitive aircraft instruments such as air speed indicators and rate of climb indicators (variometres) have connections both to the internal part of the aneroid chamber and to an external enclosing chamber.

Electronic Pressure Sensors

Piezoresistive Strain Gage: Uses the piezoresistive effect of bonded or formed strain gauges to detect strain due to applied pressure.

Capacitive: Uses a diaphragm and pressure cavity to create a variable capacitor to detect strain due to applied pressure.

Magnetic: Measures the displacement of a diaphragm by means of changes in inductance (reluctance), LVDT, Hall Effect, or by eddy current principal.

Piezoelectric: Uses the piezoelectric effect in certain materials such as quartz to measure the strain upon the sensing mechanism due to pressure.

Optical: Uses the physical change of an optical fiber to detect strain due applied pressure.

Potentiometric: Uses the motion of a wiper along a resistive mechanism to detect the strain caused by applied pressure.

Resonant: Uses the changes in resonant frequency in a sensing mechanism to measure stress, or changes in gas density, caused by applied pressure.

Thermal Conductivity

Generally, as a real gas increases in density -which may indicate an increase in pressure- its ability to conduct heat increases. In this type of gauge, a wire filament is heated by running current through it. A thermocouple or Resistance Temperature Detector (RTD) can then be used to measure the temperature of the filament. This temperature is dependent on the rate at which the filament loses heat to the surrounding gas, and therefore on the thermal conductivity. A common variant is the Pirani gauge, which uses a single platinum filament as both the heated element and RTD. These gauges are accurate from 10 Torr to 10^{-3} Torr, but they are sensitive to the chemical composition of the gases being measured.

Two-Wire

One wire coil is used as a heater, and the other is used to measure nearby temperature due to convection.

Pirani (One Wire)

A Pirani gauge consist of a metal wire open to the pressure being measured. The wire is heated by a current flowing through it and cooled by the gas surrounding it. If the gas pressure is reduced, the cooling effect will decrease, hence the equilibrium temperature of the wire will increase. The resistance of the wire is a function of its temperature: by measuring the voltage across the wire and the current flowing through it, the resistance (and so the gas pressure) can be determined. This type of gauge was invented by Marcello Pirani.

Thermocouple gauges and thermistor gauges work in a similar manner, except a thermocouple or thermistor is used to measure the temperature of the wire.

Useful range: 10^{-3} - 10 Torr (roughly 10^{-1} - 1000 Pa)

Ionization Gauge

Ionization gauges are the most sensitive gauges for very low pressures (also referred to as hard or high vacuum). They sense pressure indirectly by measuring the electrical ions produced when the gas is bombarded with electrons. Fewer ions will be produced by lower density gases. The calibration of an ion gauge is unstable and dependent on the nature of the gases being measured, which is not always known. They can be calibrated against a McLeod gauge which is much more stable and independent of gas chemistry.

Thermionic emission generate electrons, which collide with gas atoms and generate positive ions. The ions are attracted to a suitably biased electrode known as the collector. The current in the collector is proportional to the rate of ionization, which is a function of the pressure in the system. Hence, measuring the collector current gives the gas pressure. There are several sub-types of ionization gauge.

Useful range: 10^{-10} - 10^{-3} torr (roughly 10^{-8} - 10^{-1} Pa)

Most ion gauges come in two types: hot cathode and cold cathode. A third type that is more sensitive and expensive known as a spinning rotor gauge exists, but is not discussed here. In the hot cathode version, an electrically heated filament produces an electron beam. The electrons travel through the gauge and ionize gas molecules around them. The resulting ions are collected at a negative electrode. The current depends on the number of ions, which depends on the pressure in the gauge. Hot cathode gauges are accurate from 10^{-3} Torr to 10^{-10} Torr. The principle behind cold cathode version is the same, except that electrons are produced in the discharge of a high voltage. Cold Cathode gauges are accurate from 10^{-2} Torr to 10^{-9} Torr. Ionization gauge calibration is very sensitive to construction geometry, chemical composition of gases being measured, corrosion and surface deposits. Their calibration can be invalidated by activation at atmospheric pressure or low vacuum. The composition of gases at high vacuums will usually be unpredictable, so a mass spectrometre must be used in conjunction with the ionization gauge for accurate measurement.

Hot Cathode

A hot-cathode ionization gauge is composed mainly of three electrodes acting together as a triode, wherein the cathode is the filament. The three electrodes are a collector or plate, a filament, and a grid. The collector current is measured in picoamps by an electrometre. The filament voltage to ground is usually at a potential of 30 volts, while the grid voltage at 180–210 volts DC, unless there is an optional electron bombardment feature, by heating the grid, which may have a high potential of approximately 565 volts. The most common ion gauge is the hot-cathode Bayard-Alpert gauge, with a small ion collector inside the grid. A glass envelope with an opening to the vacuum can surround the electrodes, but usually the Nude Gauge is inserted in the vacuum chamber directly, the pins being fed through a ceramic plate in the wall of the chamber. Hot-cathode gauges can be damaged or lose their calibration if they are exposed to atmospheric pressure or even low vacuum while hot. The measurements of a hot-cathode ionization gauge are always logarithmic.

Electrons emitted from the filament move several times in back and forth movements around the grid before finally entering the grid. During these movements, some electrons collide with a gaseous molecule to form a pair of an ion and an electron (Electron ionization). The number of these ions is proportional to the gaseous molecule density multiplied by the electron current emitted from the filament, and these ions pour into the collector to form an ion current. Since the gaseous molecule density is proportional to the pressure, the pressure is estimated by measuring the ion current.

The low-pressure sensitivity of hot-cathode gauges is limited by the photoelectric effect. Electrons hitting the grid produce x-rays that produce photoelectric noise in the ion collector. This limits the range of older hot-cathode gauges to 10^{-8} Torr and the Bayard-Alpert to about 10^{-10} Torr. Additional wires at cathode potential in the line of sight between the ion collector and the grid prevent this effect. In the extraction type the ions are not attracted by a wire, but by an open cone. As the ions cannot decide which part of the cone to hit, they pass through the hole and form an ion beam. This ion beam can be passed on to a:

- Faraday cup
- Microchannel plate detector with Faraday cup
- Quadrupole mass analyzer with Faraday cup
- Quadrupole mass analyzer with Microchannel plate detector Faraday cup
- ion lens and acceleration voltage and directed at a target to form a sputter gun. In this case a valve lets gas into the grid-cage.

Cold Cathode

There are two subtypes of cold-cathode ionization gauges: the Penning gauge (invented by Frans Michel Penning), and the Inverted magnetron, also called a Redhead gauge. The major difference between the two is the position of the anode with respect to the cathode. Neither has a filament, and each may require a DC potential of about 4 kV for operation. Inverted magnetrons can measure down to $1x10^{-12}$ Torr.

Likewise, cold-cathode gauges may be reluctant to start at very low pressures, in that the near-absence of a gas makes it difficult to establish an electrode current - in particular in Penning gauges, which use an axially symmetric magnetic field to create path lengths for electrons that are of the order of metres. In ambient air, suitable ion-pairs are ubiquitously formed by cosmic radiation; in a Penning gauge, design features are used to ease the set-up of a discharge path. For example, the electrode of a Penning gauge is usually finely tapered to facilitate the field emission of electrons. Maintenance cycles of cold cathode gauges are, in general, measured in years, depending on the gas type and pressure that they are operated in. Using a cold cathode gauge in gases with substantial organic components, such as pump oil fractions, can result in the growth of delicate carbon films and shards within the gauge that eventually either short-circuit the electrodes of the gauge or impede the generation of a discharge path.

Calibration

Pressure gauges are either direct- or indirect-reading. Hydrostatic and elastic gauges measure pressure are directly influenced by force exerted on the surface by incident particle flux, and are called direct reading gauges. Thermal and ionization gauges read pressure indirectly by measuring a gas property that changes in a predictable manner with gas density. Indirect measurements are susceptible to more errors than direct measurements.

- Dead-weight tester
- McLeod
- mass spec + ionization

Dynamic Transients

When fluid flows are not in equilibrium, local pressures may be higher or lower than the average pressure in a medium. These disturbances propagate from their source as longitudinal pressure variations along the path of propagation. This is also called sound. Sound pressure is the instantaneous local pressure deviation from the average pressure caused by a sound wave. Sound pressure can be measured using a microphone in air and a hydrophone in water. The effective sound pressure is the root mean square of the instantaneous sound pressure over a given interval of time. Sound pressures are normally small and are often expressed in units of microbar.

- frequency response of pressure sensors
- resonance

7

Flow Measurement

Flow measurement is the quantification of bulk fluid movement. Flow can be measured in a variety of ways. Positive-displacement flow metres accumulate a fixed volume of fluid and then count the number of times the volume is filled to measure flow. Other flow measurement methods rely on forces produced by the flowing stream as it overcomes a known constriction, to indirectly calculate flow. Flow may be measured by measuring the velocity of fluid over a known area.

Units of Measurement

Both gas and liquid flow can be measured in volumetric or mass flow rates, such as litres per second or kilograms per second. These measurements can be converted between one another if the material's density is known. The density for a liquid is almost independent of the liquid conditions; however, this is not the case for gas, the density of which depends greatly upon pressure, temperature and to a lesser extent, the gas composition.

When gases or liquids are transferred for their energy content, such as the sale of natural gas, the flow rate may also be expressed in terms of energy flow, such as GJ/hour or BTU/day. The energy flow rate is the volume flow rate multiplied by the energy content per unit volume or mass flow rate multiplied by the energy content per unit mass. Where accurate energy comes to the time of the legit flow rate is desired, most flow metres will be used to calculate the volume or mass flow rate which is then adjusted to the energy flow rate by the use of a flow computer.

In engineering contexts, the volumetric flow rate is usually given the symbol Q, and the mass flow rate, the symbol $\dot{m}$.

Gas

Gases are compressible and change volume when placed under pressure, are heated or are cooled. A volume of gas under one set of pressure and temperature conditions is not equivalent to the same gas under different conditions. References will be made to "actual" flow rate through a metre and "standard" or "base" flow rate through a metre with units such as *acm/h* (actual cubic metres per hour), *kscm/h* (thousand standard cubic metres per hour), LFM (linear feet per minute), or *MSCFD* (million standard cubic feet per day).

Gas *mass* flow rate can be directly measured, independent of pressure and temperature effects, with thermal mass flow metres, Coriolis mass flow metres, or mass flow controllers.

Liquid

For liquids, various units are used depending upon the application and industry, but might include gallons (U.S. liquid or imperial) per minute, litres per second, bushels per minute or, when describing river flows, cumecs (cubic metres per second) or acre-feet per day. In oceanography a common unit to measure volume transport (volume of water transported by a current for example) is a sverdrup (Sv) equivalent to 10^6 m^3 / s.

Mechanical Flow Metres

A bucket and a stopwatch is an analogy for the operation of a positive displacement metre. The stopwatch is started when the flow starts, and stopped when the bucket reaches its limit. The volume divided by the time gives the flow rate. For continuous measurements, we need a system of continually filling and emptying buckets to divide the flow without letting it out of the pipe. These continuously forming and collapsing volumetric displacements may take the form of pistons reciprocating in cylinders, gear teeth mating against the internal wall of a metre or through a progressive cavity created by rotating oval gears or a helical screw.

Piston Metre/Rotary Piston

Because they are used for domestic water measurement, piston metres, also known as rotary piston or semi-positive displacement metres, are the most common flow measurement devices in the UK and are used for almost all metre sizes up to and including 40 mm (1½°). The piston metre operates on the principle of a piston rotating within a chamber of known volume. For each rotation, an amount of water passes through the piston chamber. Through a gear mechanism and, sometimes, a magnetic drive, a needle dial and odometre type display are advanced.

Gear Metre

Oval gear metre: An oval gear metre is a positive displacement metre that uses two or more oblong gears configured to rotate at right angles to one another, forming a T shape. Such a metre has two sides, which can be called A and B. No fluid passes through the centre of the metre, where the teeth of the two gears always mesh. On one side of the metre (A), the teeth of the gears close off the fluid flow because the elongated gear on side A is protruding into the measurement chamber, while on the other side of the metre (B), a cavity holds a fixed volume of fluid in a measurement chamber. As the fluid pushes the gears, it rotates them, allowing the fluid in the measurement chamber on side B to be released into the outlet port. Meanwhile, fluid entering the inlet port will be driven into the measurement chamber of side A, which is now open. The teeth on side B will now close off the fluid from entering side B. This cycle continues as the gears rotate and fluid is metred through alternating measurement chambers. Permanent magnets in the rotating gears can transmit a signal to an electric reed switch or current transducer for flow measurement.

Helical Gear

Helical gear flow metres get their name from the shape of their gears or rotors. These rotors resemble the shape of a helix, which is a spiral-shaped structure. As the fluid flows through the metre, it enters the compartments in the rotors, causing the rotors to rotate. Flowrate is calculated from the speed of rotation.

Nutating Disk Metre

This is the most commonly used measurement system for measuring water supply. The fluid, most commonly water, enters in one side of the metre and strikes the nutating disk, which is eccentrically mounted. The disk must then "wobble" or nutate about the vertical axis, since the bottom and the top of the disk remain in contact with the mounting chamber. A partition separates the inlet and outlet chambers. As the disk nutates, it gives direct indication of the volume of the liquid that has passed through the metre as volumetric flow is indicated by a gearing and register arrangement, which is connected to the disk. It is reliable for flow measurements within 1 percent.

Variable Area Metre

The variable area (VA) metre, also commonly called a rotametre, consists of a tapered tube, typically made of glass, with a float inside that is pushed up by fluid flow and pulled down by gravity. As flow rate increases, greater viscous and pressure forces on the float cause it to rise until it becomes stationary at a location in the tube that is wide enough for the forces to balance. Floats are made in many different shapes, with spheres and spherical ellipses being the most common. Some are designed to spin visibly in the fluid stream to aid the user in determining whether the float is stuck or not. Rotametres are available for a wide range of liquids but are most commonly used with water or air. They can be made to reliably measure flow down to 1% accuracy.

Turbine Flow Metre

The turbine flow metre (better described as an axial turbine) translates the mechanical action of the turbine rotating in the liquid flow around an axis into a user-readable rate of flow (gpm, lpm, etc.). The turbine tends to have all the flow travelling around it.

The turbine wheel is set in the path of a fluid stream. The flowing fluid impinges on the turbine blades, imparting a force to the blade surface and setting the rotor in motion. When a steady rotation speed has been reached, the speed is proportional to fluid velocity.

Turbine flow metres are used for the measurement of natural gas and liquid flow. Turbine metres are less accurate than displacement and jet metres at low flow rates, but the measuring element does not occupy or severely restrict the entire path of flow. The flow direction is generally straight through the metre, allowing for higher flow rates and less pressure loss than displacement-type metres. They are the metre of choice for large commercial users, fire protection, and as master metres for the water distribution system. Strainers are generally required to be installed in front of the metre to protect the measuring element from gravel or other debris that could enter the water distribution system. Turbine metres are generally available for 1-1/ 2" to 12" or higher pipe sizes. Turbine metre bodies are commonly made of bronze, cast Iron, or ductile iron. Internal turbine elements can be plastic or non-corrosive metal alloys. They are accurate in normal working conditions to 0.2l/s however are affect greatly with dog mix interference.

Fire metres are a specialized type of turbine metre with approvals for the high flow rates required in fire protection systems. They are often approved by Underwriters Laboratories (UL) or Factory Mutual (FM) or similar authorities for use in fire protection. Portable turbine metres may be temporarily installed to measure water used from a fire hydrant. The metres are normally made of aluminium to be light weight, and are usually 3" capacity. Water utilities often require them for measurement of water used in construction, pool filling, or where a permanent metre is not yet installed.

Woltmann Metre

The Woltmann metre comprises a rotor with helical blades inserted axially in the flow, much like a ducted fan; it can be considered a type of turbine flow metre. They are commonly referred to as helix metres, and are popular at larger sizes.

Single Jet Metre

A single jet metre consists of a simple impeller with radial vanes, impinged upon by a single jet. They are increasing in popularity in the UK at larger sizes and are commonplace in the EU.

Paddle Wheel Metre

This is similar to the single jet metre, except that the impeller is small with respect to the width of the pipe, and projects only partially into the flow, like the paddle wheel on a Mississippi riverboat.

Multiple Jet Metre

A multiple jet or multijet metre is a velocity type metre which has an impeller which rotates horizontally on a vertical shaft. The impeller element is in a housing in which multiple inlet ports direct the fluid flow at the impeller causing it to rotate in a specific direction in proportion to the flow velocity. This metre works mechanically much like a single jet metre except that the ports direct the flow at the impeller equally from several points around the circumference of the element, not just one point; this minimizes uneven wear on the impeller and its shaft.

Pelton Wheel

The Pelton wheel turbine (better described as a radial turbine) translates the mechanical action of the Pelton wheel rotating in the liquid flow around an axis into a user-readable rate of flow (gpm, lpm, etc.). The Pelton wheel tends to have all the flow travelling around it with the inlet flow focused on the blades by a jet. The original Pelton wheels were used for the generation of power and consisted of a radial flow turbine with "reaction cups" which not only move with the force of the water on the face but return the flow in opposite direction using this change of fluid direction to further increase the efficiency of the turbine.

Current Metre

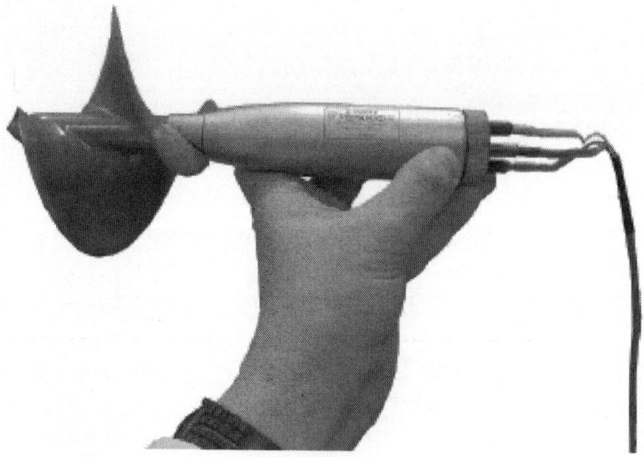

Figure: *A propeller-type current metre as used for hydroelectric turbine testing.*

Flow through a large penstock such as used at a hydroelectric power plant can be measured by averaging the flow velocity over the entire area. Propeller-type current metres (similar to the purely mechanical Ekman current metre, but now with electronic data acquisition) can be traversed over the area of the penstock and velocities averaged to calculate total flow. This may be on the order of hundreds of cubic metres per second. The flow must be kept steady during the traverse of the current metres. Methods for testing hydroelectric turbines are given in IEC standard 41. Such flow measurements are often commercially important when testing the efficiency of large turbines.

Pressure-based Metres

There are several types of flow metre that rely on Bernoulli's principle, either by measuring the differential pressure within a constriction, or by measuring static and stagnation pressures to derive the dynamic pressure.

Venturi Metre

A Venturi metre constricts the flow in some fashion, and pressure sensors measure the differential pressure before and within the constriction. This method is widely used to measure flow rate in the transmission of gas through pipelines, and has been used since Roman Empire times. The coefficient of discharge of Venturi metre ranges from 0.93 to 0.97. The first large-scale Venturi metres to measure liquid flows were developed by Clemens Herschel who used them to measure small and large flows of water and wastewater beginning at the end of the 19th century.

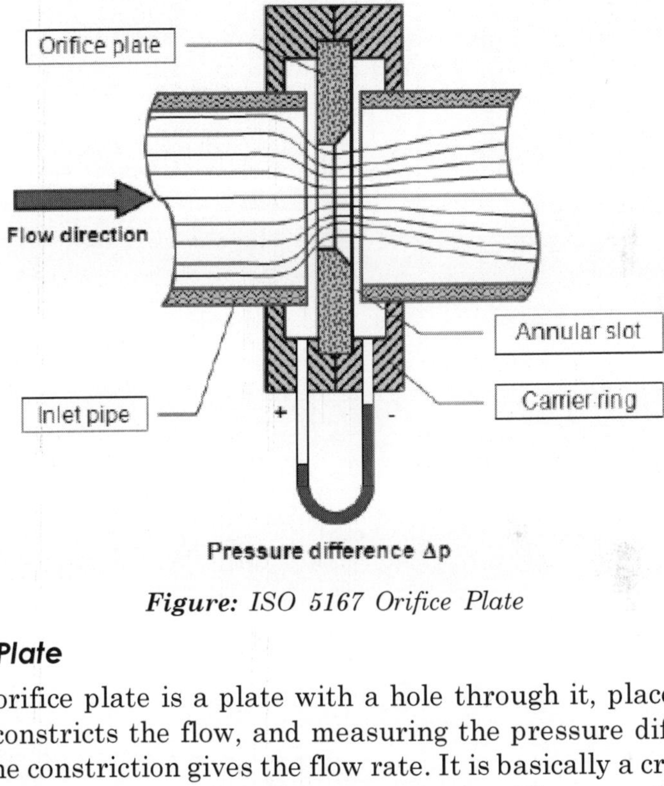

Figure: *ISO 5167 Orifice Plate*

Orifice Plate

An orifice plate is a plate with a hole through it, placed in the flow; it constricts the flow, and measuring the pressure differential across the constriction gives the flow rate. It is basically a crude form of Venturi metre, but with higher energy losses. There are three type of orifice: concentric, eccentric, and segmental.

Dall Tube

The Dall tube is a shortened version of a Venturi metre, with a lower pressure drop than an orifice plate. As with these flow metres the flow rate in a Dall tube is determined by measuring the pressure drop caused by restriction in the conduit. The pressure differential is typically measured using diaphragm pressure transducers with digital readout. Since these metres have significantly lower permanent pressure losses than orifice metres, Dall tubes are widely used for measuring the flow rate of large pipeworks.

Pitot Tube

A Pitot tube is a pressure measuring instrument used to measure fluid flow velocity by determining the stagnation pressure. Bernoulli's equation is used to calculate the dynamic pressure and hence fluid velocity.

Multi-Hole Pressure Probe

Multi-hole pressure probes (also called impact probes) extend the theory of pitot tube to more than one dimension. A typical impact probe consists of three or more holes (depending on the type of probe) on the measuring tip arranged in a specific pattern. More holes allow the instrument to measure the direction of the flow velocity in addition to its magnitude (after appropriate calibration). Three holes arranged in a line allow the pressure probes to measure the velocity vector in two dimensions. Introduction of more holes, e.g. five holes arranged in a "plus" formation, allow measurement of the three-dimensional velocity vector.

Cone Metres

Figure: *8inch (200mm) V-Cone Flowmeter shown with ANSI 300# raised face weld neck flanges*

Cone metres are a newer differential pressure metering device first launched in 1985 by McCrometer in Hemet, CA. While working with the same basic principles as Ventrui and Orifice type DP metres, cone metres don't require the same upstream and downstream piping. The cone acts as a conditioning device as well as a differential pressure producer. Upstream requirements are between 0-5 diameters compared to up to 44 diameters for an orifice plate or 22 diameters for a Venturi. Because cone metres are generally of welded construction, it is recommended they are always calibrated prior to service. Inevitably heat effects of welding cause distortions and other effects that prevent tabular data on discharge coefficients with respect to line size, beta ratio and operating Reynolds Numbers from being collected and published. Calibrated cone metres have an uncertainty up to +/-0.5%. Un-calibrated cone metres have an uncertainty of +/-5.0%.

Optical Flow Metres

Optical flow metres use light to determine flow rate. Small particles which accompany natural and industrial gases pass through two laser beams focused a short distance apart in the flow path, in a pipe by illuminating optics. Laser light is scattered when a particle crosses the first beam. The detecting optics collects scattered light on a photodetector, which then generates a pulse signal. As the same particle crosses the second beam, the detecting optics collect scattered light on a second photodetector, which converts the incoming light into a second electrical pulse. By measuring the time interval between these pulses, the gas velocity is calculated as $V = D / t$ where D is the distance between the laser beams and t is the time interval.

Laser-based optical flow metres measure the actual speed of particles, a property which is not dependent on thermal conductivity of gases, variations in gas flow or composition of gases. The operating principle enables optical laser technology to deliver highly accurate flow data, even in challenging environments which may include high temperature, low flow rates, high pressure, high humidity, pipe vibration and acoustic noise.

Optical flow metres are very stable with no moving parts and deliver a highly repeatable measurement over the life of the product. Because distance between the two laser sheets does not change, optical flow metres do not require periodic calibration after their initial commissioning. Optical flow metres require only one installation point, instead of the two installation points typically required by other types of metres. A single installation point is simpler, requires less maintenance and is less prone to errors. Commercially available optical flow metres are capable of measuring flow from 0.1 m/s to faster than 100 m/s (1000:1 turn down ratio) and have been demonstrated to be effective for the measurement of flare gases from oil wells and refineries, a contributor to atmospheric pollution.

Open Channel Flow Measurement

Level to flow: The level of the water is measured at a designated point behind weir or in flume a hydraulic structure using various secondary devices (bubblers, ultrasonic, float, and differential pressure are common methods). This depth is converted to a flow rate according to a theoretical formula of the form $Q = KH^X$ where Q is the flow rate, K is a constant, H is the water level, and X is an exponent which varies with the device used; or it is converted according to empirically derived level/flow data points (a "flow curve"). The flow rate can then integrated over time into volumetric flow. Level to flow devices are commonly used to measure the flow of surface waters (springs, stream, and rivers), industrial discharges, and sewage. Of these, weirs are used on flow streams with low solids (typically surface waters), while flumes are used on flows containing low or high solids contents.

Area/Velocity

The cross-sectional area of the flow is calculated from a depth measurement and the average velocity of the flow is measured directly (Doppler and propeller methods are common). Velocity times the cross-sectional area yields a flow rate which can be integrated into volumetric flow.

Dye Testing

A known amount of dye (or salt) per unit time is added to a flow stream. After complete mixing, the concentration is measured. The dilution rate equals the flow rates.

Acoustic Doppler Velocimetry

Acoustic Doppler velocimetry (ADV) is designed to record instantaneous velocity components at a single point with a relatively high frequency. Measurements are performed by measuring the velocity of particles in a remote sampling volume based upon the Doppler shift effect.

Thermal Mass Flow Metres

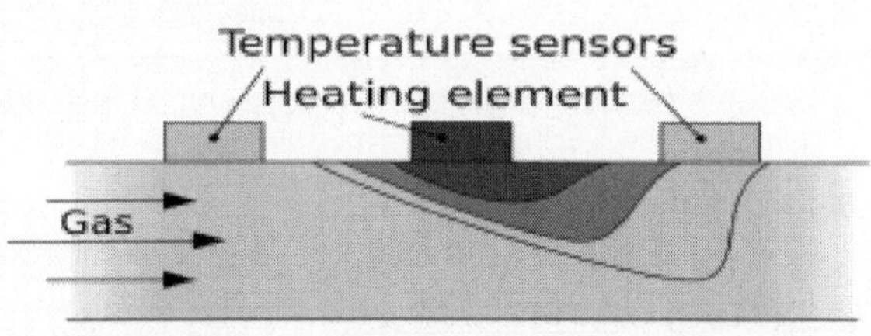

Figure: *Temperature at the sensors varies depending upon the mass flow*

Thermal mass flow metres generally use combinations of heated elements and temperature sensors to measure the difference between static and flowing heat transfer to a fluid and infer its flow with a knowledge of the fluid's specific heat and density. The fluid temperature is also measured and compensated for. If the density and specific heat characteristics of the fluid are constant, the metre can provide a direct mass flow readout, and does not need any additional pressure temperature compensation over their specified range.

Technological progress has allowed the manufacture of thermal mass flow metres on a microscopic scale as MEMS sensors; these flow devices can be used to measure flow rates in the range of nanolitres or microlitres per minute.

Thermal mass flow metre (also called thermal dispersion flowmetre) technology is used for compressed air, nitrogen, helium, argon, oxygen, and natural gas. In fact, most gases can be measured as long as they are fairly clean and non-corrosive. For more aggressive gases, the metre may be made out of special alloys (e.g. Hastelloy), and pre-drying the gas also helps to minimize corrosion.

Today, thermal mass flow metres are used to measure the flow of gases in a growing range of applications, such as chemical reactions or thermal transfer applications that are difficult for other flow metreing technologies. This is because thermal mass flow metres monitor variations in one or more of the thermal characteristics (temperature, thermal conductivity, and/or specific heat) of gaseous media to define the mass flow rate.

The MAF Sensor

In many late model automobiles, a mass airflow sensor (MAF sensor) is used to accurately determine the mass flowrate of intake air used in the internal combustion engine. Many such mass flow sensors utilise a heated element and a downstream temperature sensor to indicate the air flowrate. Other sensors use a spring-loaded vane. In either case, the vehicle's electronic control unit interprets the sensor signals as a real time indication of an engine's fuel requirement.

Vortex Flow Metres

Another method of flow measurement involves placing a bluff body (called a shedder bar) in the path of the fluid. As the fluid passes this bar, disturbances in the flow called vortices are created. The vortices trail behind the cylinder, alternatively from each side of the bluff body. This vortex trail is called the Von Kármán vortex street after von Kármán's 1912 mathematical description of the phenomenon. The frequency at which these vortices alternate sides is essentially proportional to the flow rate of the fluid. Inside, atop, or downstream of the shedder bar is a sensor for measuring the frequency of the vortex shedding. This sensor is often a piezoelectric crystal, which produces a small, but measurable, voltage pulse every time a vortex is created. Since the frequency of such a voltage pulse is also proportional to the fluid velocity, a volumetric flow rate is calculated using the cross sectional area of the flow metre. The frequency is measured and the flow rate is calculated by the flowmetre electronics using the equation $f = SV / L$ where f is the frequency of the vortices, L the characteristic length of the bluff body, V is the velocity of the flow over the bluff body, and S is the Strouhal number, which is essentially a constant for a given body shape within its operating limits.

Electromagnetic, Ultrasonic and Coriolis Flow Metres

Modern innovations in the measurement of flow rate incorporate electronic devices that can correct for varying pressure and temperature (i.e. density) conditions, non-linearities, and for the characteristics of the fluid.

Magnetic Flow Metres

Figure: *A magnetic flow meter at the Tetley's Brewery in Leeds, West Yorkshire.*

Magnetic flow metres, often called “mag metre”s or “electromag”s, use a magnetic field applied to the metreing tube, which results in a potential difference proportional to the flow velocity perpendicular to the flux lines. The potential difference is sensed by electrodes aligned perpendicular to the flow and the applied magnetic field. The physical principle at work is Faraday’s law of electromagnetic induction. The magnetic flow metre requires a conducting fluid and a nonconducting pipe liner. The electrodes must not corrode in contact with the process fluid; some magnetic flowmetres have auxiliary transducers installed to clean the electrodes in place. The applied magnetic field is pulsed, which allows the flowmetre to cancel out the effect of stray voltage in the piping system.

Ultrasonic Flow Metres (Doppler, Transit Time)

There are two main types of Ultrasonic flow metres: Doppler and transit time. While they both utilise ultrasound to make measurements and can be non-invasive (measure flow from outside the tube, pipe or vessel), they measure flow by very different methods.

Ultrasonic transit time flow metres measure the difference of the transit time of ultrasonic pulses propagating in and against the direction of flow. This time difference is a measure for the average velocity of the fluid along the path of the ultrasonic beam. By using the absolute transit times both the averaged fluid velocity and the speed of sound can be calculated. Using the two transit times t_{up} and t_{down} and the distance between receiving and transmitting transducers L and the inclination angle α one can write the equations:

$$v = \frac{L}{2\sin(\alpha)} \frac{t_{up} - t_{down}}{t_{up}\, t_{down}} \quad \text{and} \quad c = \frac{L}{2} \frac{t_{up} + t_{down}}{t_{up}\, t_{down}}$$

where v is the average velocity of the fluid along the sound path and c is the speed of sound.

With wide-beam illumination transit time ultrasound can also be used to measure volume flow independent of the cross-sectional area of the vessel or tube.

Ultrasonic Doppler flow metres measure the Doppler shift resulting from reflecting an ultrasonic beam off the particulates in flowing fluid. The frequency of the transmitted beam is affected by the movement of the particles; this frequency shift can be used to calculate the fluid velocity. For the Doppler principle to work there must be a high enough density of sonically reflective materials such as solid particles or air bubbles suspended in the fluid. This is in direct contrast to an ultrasonic transit time flow metre, where bubbles and solid particles reduce the accuracy of the measurement. Due to the dependency on these particles there are limited applications for Doppler flow metres. This technology is also known as acoustic Doppler velocimetry.

One advantage of ultrasonic flow metres is that they can effectively measure the flow rates for a wide variety of fluids, as long as the speed of sound through that fluid is known. For example, ultrasonic flow metres are used for the measurement of such diverse fluids a liquid natural gas (LNG) and blood. One can also calculate the expected speed of sound for a given fluid; this can be compared to the speed of sound empirically measured by an ultrasonic flow metre for the purposes of monitoring the quality of the flow metre's measurements. A drop in quality (change in the measured speed of sound) is an indication that the metre needs servicing.

Coriolis Flow Metres

Using the Coriolis effect that causes a laterally vibrating tube to distort, a direct measurement of mass flow can be obtained in a coriolis flow metre. Furthermore a direct measure of the density of the fluid is obtained. Coriolis measurement can be very accurate irrespective of the type of gas or liquid that is measured; the same measurement tube can be used for hydrogen gas and bitumen without recalibration. Coriolis flow metres can be used for the measurement of natural gas flow.

Laser Doppler Flow Measurement

A beam of laser light impinging on a moving particle will be partially scattered with a change in wavelength proportional to the particle's speed (the Doppler effect). A Laser Doppler velocimetre (LDV), also called a laser Doppler anemometre (LDA), focuses a laser beam into a small volume in a flowing fluid containing small particles (naturally occurring or induced). The particles scatter the light with a Doppler shift. Analysis of this shifted wavelength can be used to directly, and with great precision, determine the speed of the particle and thus a close approximation of the fluid velocity.

A number of different techniques and device configurations are available for determining the Doppler shift. All use a photodetector (typically an avalanche photodiode) to convert the light into an electrical waveform for analysis. In most devices, the original laser light is divided into two beams. In one general LDV class, the two beams are made to intersect at their focal points where they interfere and generate a set of straight fringes. The sensor is then aligned to the flow such that the fringes are perpendicular to the flow direction. As particles pass through the fringes, the Doppler-shifted light is collected into the photodetector. In another general LDV class, one beam is used as a reference and the other is Doppler-scattered. Both beams are then collected onto the photodetector where optical heterodyne detection is used to extract the Doppler signal.

Calibration

Even though ideally the flowmetre should be unaffected by its environment, in practice this is unlikely to be the case. Often measurement errors originate from incorrect installation or other environment dependent factors. In situ methods are used when flow metre is calibrated in the correct flow conditions.

Transit Time Method

For pipe flows a so-called transit time method is applied where a radiotracer is injected as a pulse into the measured flow. The transit time is defined with the help of radiation detectors placed on the outside of the pipe. The volume flow is obtained by multiplying the measured average fluid flow velocity by the inner pipe cross section. This reference flow value is compared with the simultaneous flow value given by the flow measurement to be calibrated. The procedure is standardised (ISO 2975/VII for liquids and BS 5857-2.4 for gases). The best accredited measurement uncertainty for liquids and gases is 0.5%.

Tracer Dilution Method

The radiotracer dilution method is used to calibrate open channel flow measurements. A solution with a known tracer concentration is injected at a constant known velocity into the channel flow. Downstream where the tracer solution is thoroughly mixed over the flow cross section, a continuous sample is taken and its tracer concentration in relation to that of the injected solution is determined. The flow reference value is determined by using the tracer balance condition between the injected tracer flow and the diluting flow..

The procedure is standardised (ISO 9555-1 and ISO 9555-2 for liquid flow in open channels). The best accredited measurement uncertainty is 1%.

Pipe Network Analysis

In fluid dynamics, pipe network analysis is the analysis of the fluid flow through a hydraulics network, containing several or many interconnected branches. The aim is to determine the flow rates and pressure drops in the individual sections of the network. This is a common problem in hydraulic design.

Description

In order to direct water to many individuals in a municipal water supply, many times the water is routed through a water supply network. A major part of this network may consist of interconnected pipes. This network creates a special class of problems in hydraulic design typically referred to as pipe network analysis. The modern solution for this is to use specialized software in order to automatically solve the problems. However, the problems can also be addressed with simpler methods like a spreadsheet equipped with a solver, or a modern graphing calculator.

Network Analysis

Once the friction factors are solved for, then we can start considering the network problem. We can solve the network by satisfying two conditions.

1. At any junction, the flow into a junction equals the flow out of the junction.
2. Between any two junctions, the head loss is independent of the path taken.

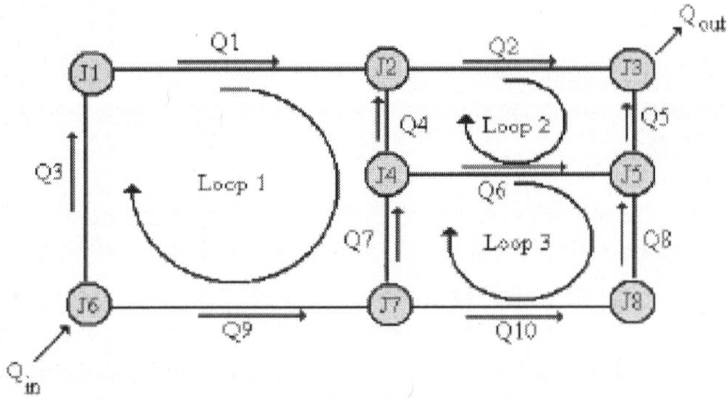

The classical approach for solving these networks is to use the Hardy Cross method. In this formulation, first you go through and create guess values for the flows in the network. That is, if Q7 enters a junction and Q6 and Q4 leave the same junction, then the initial guess must satisfy Q7 = Q6 + Q4. After the initial guess is made, then, a loop is considered so that we can evaluate our second condition. Given a starting node, we work our way around the loop in a clockwise fashion, as illustrated by Loop 1. We add up the head losses according to the Darcy–Weisbach equation for each pipe if Q is in the same direction as our loop like Q1, and subtract the head loss if the flow is in the reverse direction, like Q4. In order to satisfy the second condition, we should end up with 0 about the loop if the network is completely solved. If the actual sum of our head loss is not equal to 0, then we will adjust all the flows in the loop by an amount given by the following formula, where a positive adjustment is in the clockwise direction.

$$\Delta Q = \frac{\sum \text{head loss}_c - \sum \text{head loss}_{cc}}{n \cdot (\sum \frac{\text{head loss}_c}{Q_c} + \sum \frac{\text{head loss}_{cc}}{Q_{cc}})},$$

where

- n is 1.85 for Hazen-Williams and
- n is 2 for Darcy–Weisbach.

The clockwise specifier (c) means only the flows that are moving clockwise in our loop, while the counter-clockwise specifier (cc) is only the flows that are moving counter-clockwise.

This adjustment won't solve the problem, since with most networks we will have several loops. It is ok to do this adjustment, however, because our flow changes won't alter condition 1, and therefore, our other loops will still satisfy condition 1. However, we should use the results from the first loop if we progress to any other loops.

The more modern method is simply to create a set of conditions from your junctions and head-loss criteria. Then, use a Root-finding algorithm to find Q values that satisfy all the equations. The literal friction loss equations will use a term called Q^2, but we want to preserve any changes in direction. Create a separate equation for each loop where the head losses are added up, but instead of squaring Q, use $|Q| \cdot Q$ instead (with $|Q|$ the absolute value of Q) for the formulation so that any sign changes will reflect appropriately in the resulting head-loss calculation.

Hardy Cross Method

The Hardy Cross method is an iterative method for determining the flow in pipe network systems where the inputs and outputs are known, but the flow inside the network is unknown. The method was first published in November 1936 by its namesake, Hardy Cross, a structural engineering professor at the University of Illinois at Urbana–Champaign. The Hardy Cross method is an adaptation of the Moment distribution method, which was also developed by Hardy Cross as a way to determine the moments in indeterminate structures.

The introduction of the Hardy Cross method for analyzing pipe flow networks revolutionized municipal water supply design. Before the method was introduced, solving complex pipe systems for distribution was extremely difficult due to the nonlinear relationship between head loss and flow.

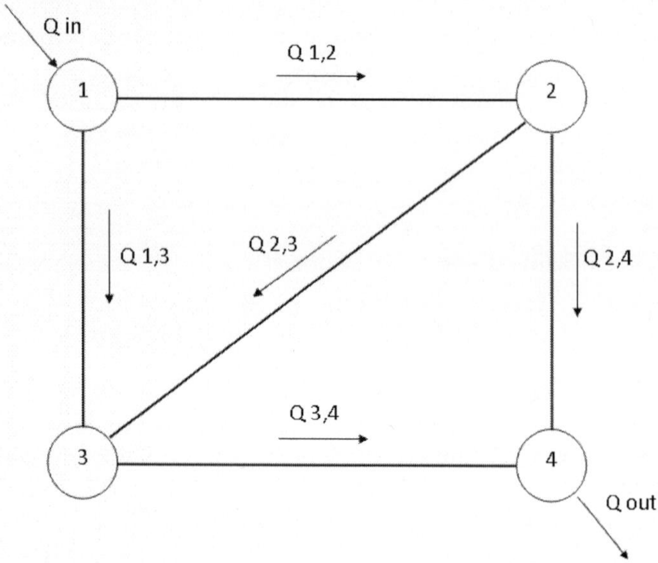

Figure: *An example pipe flow network*

The method was later made obsolete by computer solving algorithms employ the Newton-Raphson method or other solving methods that prevent the need to solve nonlinear systems of equations by hand.

History

In 1930, Hardy Cross published a paper called "Analysis of Continuous Frames by Distributing Fixed-End Moments" in which he described the moment distribution method, which would change the way engineers in the field performed structural analysis. The moment distribution method was used to determine the moment distributions in indeterminant concrete structures and allowed for engineers to safely design structures from concrete from the 1930s to 1960. In November 1936, Cross applied the same geometric method to solving pipe network flow distribution problems, and published a paper called "Analysis of flow in networks of conduits or conductors."

Derivation

The Hardy Cross method is an application of continuity of flow and continuity of potential to iteratively solve for flows in a pipe network. In the case of pipe flow, conservation of flow means that the flow in is equal to the flow out at each junction in the pipe. Conservation of potential means that the total directional head loss along any loop in the system is zero (assuming that a head loss counted against the flow is actually a head gain).

Hardy Cross developed two methods for solving flow networks. Each method starts by maintaining either continuity of flow or potential, and then iteratively solves for the other.

Assumptions

The Hardy Cross method assumes that the flow going in and out of the system is known and that the pipe length, diameter, roughness and other key characteristics are also known or can be assumed. The method also assumes that the relation between flow rate and head loss is known, but the method does not require any particular relation to be used.

In the case of water flow through pipes, a number of methods have been developed to determine the relationship between head loss and flow. The Hardy Cross method allows for any of these relationships to be used.

The general relationship between head loss and flow is:

$$h_f = r \cdot Q^n$$

where r is the head loss per unit flow and n is the flow exponent. In most design situations the values that make up r, such as pipe length, diameter, and roughness, are taken to be known or assumed and the value of r can be determined for each pipe in the network. The values that make up r and the value of n change depending on the relation used to determine head loss. However, all relations are compatible with the Hardy Cross method.

Head Loss Equation	***Relation***	***r***	***n***
Hazen-Williams equation	$h_f = L \cdot \frac{10.67}{C^{1.85}} \frac{Q^{1.85}}{d^{4.87}}$	$L \cdot \frac{10.67}{C^{1.85} \quad d^{4.87}}$	1.85
Darcy-Weisbach equation	$h_f = \frac{8fLQ^2}{g\pi^2 D^5}$	$\frac{8fL}{g\pi^2 D^5}$	2

It is also worth noting that the Hardy Cross method can be used to solve simple circuits and other flow like situations. In the case of simple circuits,

$$V = R \cdot I$$

is equivalent to

$$h_f = r \cdot Q^n .$$

By setting the coefficient r to R, the flow rate Q to I and the exponent n to 1, the Hardy Cross method can be used to solve a simple circuit. However, because the relation between the voltage drop and current is linear, the Hardy Cross method is not necessary and the circuit can be solved using non-iterative methods.

Method of Balancing Heads

The method of balancing heads uses an initial guess that satisfies continuity of flow at each junction and then balances the flows until continuity of potential is also achieved over each loop in the system.

Proof

The following proof is taken from Hardy Cross's paper, "Analysis of flow in networks of conduits or conductors.", and can be verified by National Programme on Technology Enhanced Learning Water and Wastewater Engineering page, and Fundamentals of Hydraulic Engineering Systems by Robert J. Houghtalen.

If the initial guess of flow rates in each pipe is correct, the change in head over a loop in the system, ΣrQ^n would be equal to zero. However, if the initial guess is not correct, then the change in head will be non-zero and a change in flow, ΔQ must be applied. The new flow rate, $Q = Q_0 + \Delta Q$ is the sum of the old flow rate and some change in flow rate such that the changed in head over the loop is zero. The sum of the change in head over the new loop will then be $\Sigma r(Q_0 + \Delta Q)^n = 0$. The value of $\Sigma r(Q_0 + \Delta Q)^n$ can be approximated using the Taylor expansion.

$$\Sigma r(Q_0 + \Delta Q)^n = \Sigma r(Q_0^n + nQ_0^{n-1}\Delta Q + \ldots) = 0$$

For a small ΔQ compared to Q_0 the additional terms vanish, leaving:

$$\Sigma r(Q_0^n + nQ_0^{n-1}\Delta Q) = 0$$

And solving for ΔQ

$$\Sigma rQ_0^n = -\Sigma nrQ_0^{n-1}\Delta Q$$

$$\Delta Q = -\frac{\Sigma rQ_0^n}{\Sigma nrQ_0^{n-1}}$$

The change in flow that will balance the head over the loop is approximated by $\Delta Q = -\frac{\Sigma rQ_0^n}{\Sigma nrQ_0^{n-1}}$. However, this is only an approximation due to the terms that were ignored from the Taylor expansion. The change in head over the loop may not be zero, but it will be smaller than the initial guess. Multiple iterations of finding a new ΔQ will approximate to the correct solution.

Process

The method is as follows:

1. Guess the flows in each pipe, making sure that the *total in flow* is equal to the *total out flow* at each junction. *(The guess doesn't have to be good, but a good guess will reduce the time it takes to find the solution.)*
2. Determine each closed loop in the system
3. For each loop, determine the clockwise head losses and counter-clockwise head losses. Head loss in each pipe are calculated using $h_f = rQ^n$. Clockwise head losses are from flows in the clockwise direction and likewise for counter-clockwise.
4. Determine the total head loss in the loop, ΣrQ^n, by subtracting the counter-clockwise head loss from the clockwise head loss.
5. For each loop, find ΣnrQ^{n-1} without reference to direction (all values should be positive).
6. The change in flow is equal to $\frac{\Sigma rQ^n}{\Sigma nrQ^{n-1}}$.
7. If the change in flow is positive, apply it to all pipes of the loop in the counter-clockwise direction. If the change in flow is negative, apply it to all pipes of the loop in the clockwise direction.
8. Continue from step 3 until the change in flow is within a satisfactory range.

Method of Balancing Flows (Section Incomplete)

The method of balancing flows uses an initial guess that satisfies continuity of potential over each loop and then balances the flows until continuity of flow is also achieved at each junction.

Process

Advantages of the Hardy Cross method:

Simple Math

The Hardy Cross method is useful because it relies on only simple math, circumventing the need to solve a system of equations. Without the Hardy Cross methods, engineers would have to solve complex systems of equations with variable exponents that cannot easily be solved by hand.

Self Correcting

The Hardy Cross method iteratively corrects for the mistakes in the initial guess used to solve the problem. Subsequent mistakes in calculation are also iteratively corrected. If the method is followed correctly, the proper flow in each pipe can still be found if small mathematical errors are consistently made in the process. As long as the last few iterations are done with attention to detail, the solution will still be correct. In fact, it is possible to intentionally leave off decimals in the early iterations of the method to run the calculations faster.

Example

The Hardy Cross method can be used to calculate the flow distribution in a pipe network. Consider the example of a simple pipe flow network shown at the right. For this example, the in and out flows will be 10 litres per second. We will consider n to be 2, and the head loss per unit flow *r*, and initial flow guess for each pipe as follows:

Pipe	*Q12*	*Q13*	*Q23*	*Q24*	*Q34*
r	1	5	1	5	1
Q guess (L/s)	5	5	0	5	5

We solve the network by method of balancing heads, following the steps outlined in method process above.

1. The initial guesses are set up so that continuity of flow is maintained at each junction in the network.
2. The loops of the system are identified as loop 1-2-3 and loop 2-3-4.
3. The head losses in each pipe are determined.

Loop 1-2-3	*Q12*	*Q13*	*Q23*
Head loss = rQ^2	25	125	0
Direction	Clockwise	Counter-clockwise	Clockwise

For loop 1-2-3, the sum of the clockwise head losses is 25 and the sum of the counter-clockwise head losses is 125.

Loop 2-3-4	*Q23*	*Q24*	*Q34*
Head loss = rQ^2	0	125	25
Direction	Counter-clockwise	Clockwise	Counter-clockwise

For loop 2-3-4, the sum of the clockwise head losses is 125 and the sum of the counter-clockwise head losses is $125-25=100$ 25.

4. The total clockwise head loss in loop 1-2-3 is $25-125=-100$. The total clockwise head loss in loop 2-3-4 is .
5. The value ΣnrQ^{n-1} of is determined for each loop. It is found to be 60 in both loops (due to symmetry), as shown in the figure.
6. The change in flow is found for each loop using the $\frac{\Sigma rQ^n}{\Sigma nrQ^{n-1}}$ equation . For loop 1-2-3, the change in flow is equal to $-100/60=-1.66$ and for loop 2-3-4 the change in flow is equal to $100/60=1.66$.
7. The change in flow is applied across the loops. For loop 1-2-3, the change in flow is negative so its absolute value is applied in the clockwise direction. For loop 2-3-4, the change in flow is positive so its absolute value is applied in the counter-clockwise direction. For pipe 2-4, which is in both loops, the changes in flow are cumulative.

Pipe	***Q12***	***Q13***	***Q23***	***Q24***	***Q34***
Q (L/s)	6.66	3.33	3.33	3.33	6.66

The process then repeats from step 3 until the change in flow becomes sufficiently small or goes to zero.

3. The total lead loss in Loop 1-2-3 is

Loop 1-2-3	***Q12***	***Q13***	***Q23***
Head loss = rQ^2	44.4	55.5	11.1
Direction	Clockwise	Counter-clockwise	Clockwise

Notice that the clockwise head loss is equal to the counter-clockwise head loss. This means that the flow in this loop is balanced and the flow rates are correct. The total head loss in loop 2-3-4 will also be balanced (again due to symmetry).

Loop 2-3-4	***Q23***	***Q24***	***Q34***
Head loss = rQ^2	11.1	55.5	44.4
Direction	Counter-clockwise	Clockwise	Counter-clockwise

In this case, the method found the correct solution in one iteration. For other networks, it may take multiple iterations until the flows in the pipes are correct or approximately correct.

Pumps and Systems

In most training exercises or general discussions of pump-system interactions, great simplifications are made. The two most common examples presented are usually two pumps operating in parallel, or two pumps operating in series. Discussion rarely involves more than two pumps, and operation of a single pump against a multi-branched system is discussed even less frequently. But real installations rarely resemble such isolated and greatly simplified situations. Typically in actual practice, multiple pumps operate against multiple branches of a system, or several interconnected complex systems operate with pumps coming on and off line and valves adding multiple new branches or shutting off parts of the system.

Such multi-branched, multi-pump systems can no longer be analyzed by hand with a graph and calculator. Instead, an entire computerized network of the plant piping, tanks, and pumps that reflects proper sizes, friction, elevations, and so on needs to be carefully modeled. However, before attempting to model the complexities of the entire plant, the fundamentals of the multi-pump, multi-branch systems that are in place must first be understood. When such an examination is reviewed, it will become clear that the entire complex network is essentially a combination of the following three cases:

1. Several pumps in parallel, against a single system
2. Several pumps in series, against a single system
3. A single pump against a two-branched system, or against a multi-branched system

Below is a simplified example exercise involving three cases that were excerpted from one of our training classes. Consider the fundamentals of three such cases as basic building blocks for an eventually more complex network:

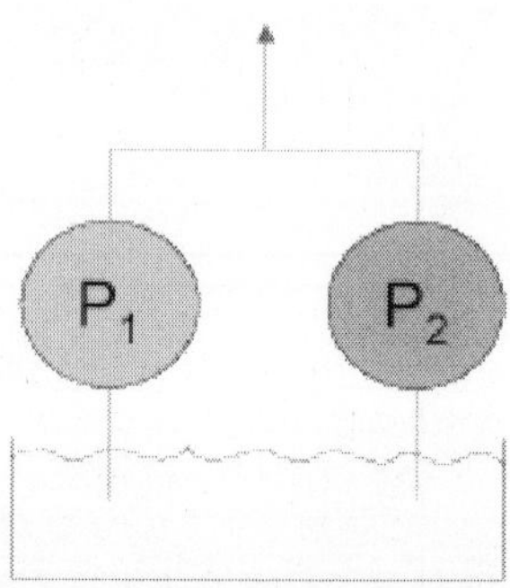

Figure: *Case (a). (2) pumps in parallel against (1) system.*

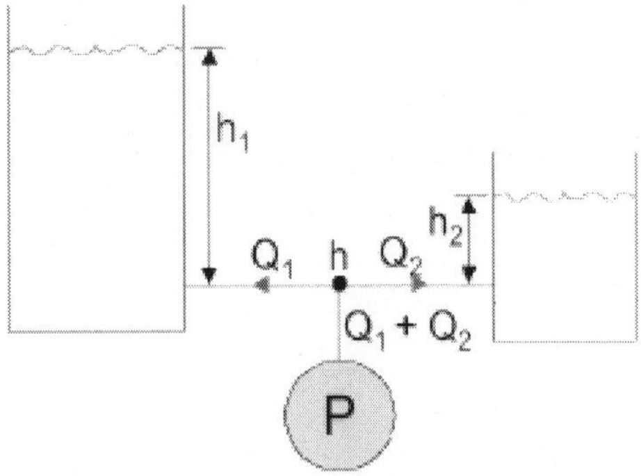

Figure: *Case (c). (1) pump - against (2) system branches.*

Cases (a) and (b) are more common, i.e. a single-branched system is considered. Case (c) is less frequently considered, but it has one pump against a multiple-branched (two in this illustration) system. Graphically, in parallel operation, (pump curves) flow is additive at constant head; in series operation, the (pump curves) head is additive at constant flow.

However, for a multiple-branched system the flow is additive (system curves) at constant head. Graphs (a), (b), and (c) below reflect the three cases above, respectively. The *operating point* is an intersection between the final pump curve and a final system curve, as shown with numerical examples below.

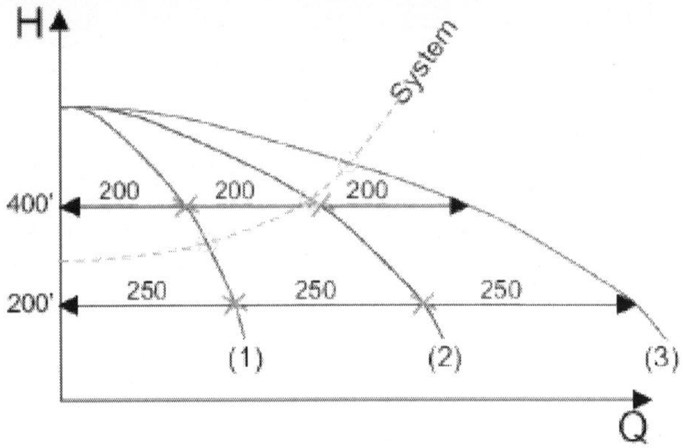

Figure: *Graph (a). (2) pumps in parallel against (1) system. Method of construction of multiple (combined) pump curve: at constant head line (200-ft), double the flow (250-gpm x 2) for a point on a 2-pump curve. Repeat at several other constant head lines; for 400-ft, double the flow (200-gpm x 2). For 3-pump operation, triple the flow at constant head lines. Continue in the same fashion for more pumps. Intersections between one, two, three, or more pumps, with a given system curve, establish the operating point (resultant head and flow) for multiple pumps.*

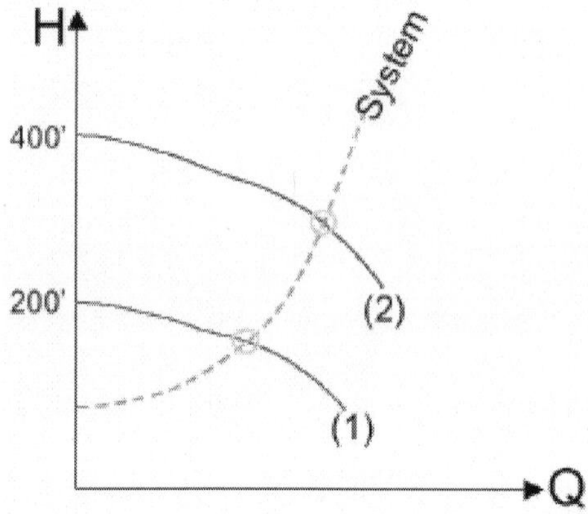

Figure: *Graph (b). (2 or more) pumps (or pump stages) in series - against (1) system. Method of construction of multiple (combined) pump curve: at series of constant flows, add pump (or pump stages for multiple pumps) heads. Intersections between the resulting pump curve (stages curve) and a system curve establishes operating point (resultant head and flow).*

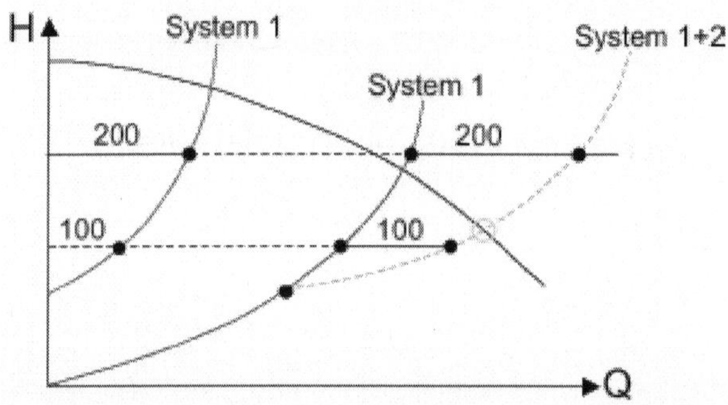

Figure: *Graph (c). (1) pump - against (2) system branches. Method of construction: instead of combining pump curves, add flows (at constant head) for systems (can be more then two). Intersection of the resultant system curve with a given pump curve produces the operating point (resultant flow and head).*

Computerizing the Process

This process can be computerized, as illustrated in a simplified example of a positive displacement pump operating against two systems (two branches). Underlying programming formulas are:

1. $H = h_1 + k_1 \times Q_1^2$ - This is the general equation for a system of branch (1), including static head h_1 and friction with system friction resistance k_1

2. $H = h_2 + k_2 \times Q_2^2$ - This is the general equation for a system of branch (2), including static head h_2 and friction with system friction resistance k_2
3. $Q = Q_1 + Q_2$ - This is what leaves the pumps splits into branches.

Known/given: Q, k_1, k_2, h_1, h_2 (where h_1 and h_2 are static heads for the two systems, and k_1 and k_2 are friction coefficients for these systems)

Need to find : Q_1, Q_2 and pump head h.

Programming procedure:

a. Guess h
b. Calculate Q_1 from (1)
c. Calculate Q_2 from (2)
d. Calculate $Q = Q_1 + Q_2$ and compare with given Q
e. If calculated flow is different than given, re-guess h and repeat the process until error is small

Boundary Layer

Figure: *Boundary layer visualization, showing transition from laminar to turbulent condition*

In physics and fluid mechanics, a boundary layer is the layer of fluid in the immediate vicinity of a bounding surface where the effects of viscosity are significant. In the Earth's atmosphere, the planetary boundary layer is the air layer near the ground affected by diurnal heat, moisture or momentum transfer to or from the surface. On an aircraft wing the boundary layer is the part of the flow close to the wing, where viscous forces distort the surrounding non-viscous flow.

Laminar boundary layers can be loosely classified according to their structure and the circumstances under which they are created. The thin shear layer which develops on an oscillating body is an example of a Stokes boundary layer, while the Blasius boundary layer refers to the well-known similarity solution near an attached flat plate held in an oncoming unidirectional flow. When a fluid rotates and viscous forces are balanced by the Coriolis effect (rather than convective inertia), an Ekman layer forms. In the theory of heat transfer, a thermal boundary layer occurs. A surface can have multiple types of boundary layer simultaneously.

Aerodynamics

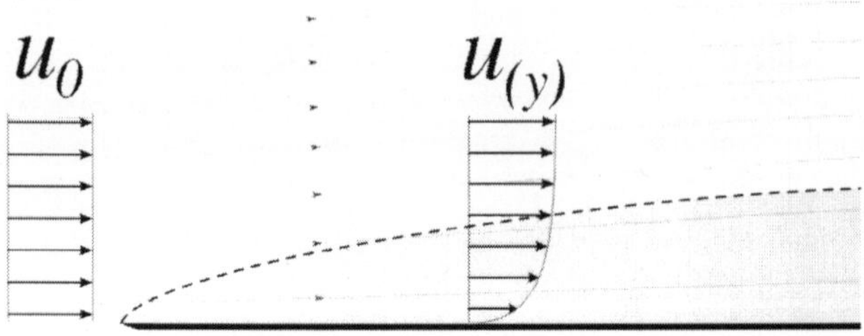

Figure: *Laminar boundary layer velocity profile*

The aerodynamic boundary layer was first defined by Ludwig Prandtl in a paper presented on August 12, 1904 at the third International Congress of Mathematicians in Heidelberg, Germany. It simplifies the equations of fluid flow by dividing the flow field into two areas: one inside the boundary layer, dominated by viscosity and creating the majority of drag experienced by the boundary body; and one outside the boundary layer, where viscosity can be neglected without significant effects on the solution. This allows a closed-form solution for the flow in both areas, a significant simplification of the full Navier–Stokes equations.

The majority of the heat transfer to and from a body also takes place within the boundary layer, again allowing the equations to be simplified in the flow field outside the boundary layer. The pressure distribution throughout the boundary layer in the direction normal to the surface (such as an airfoil) remains constant throughout the boundary layer, and is the same as on the surface itself.

The thickness of the velocity boundary layer is normally defined as the distance from the solid body at which the viscous flow velocity is 99% of the freestream velocity (the surface velocity of an inviscid flow). An alternative definition, the displacement thickness, recognises that the boundary layer represents a deficit in mass flow compared to inviscid flow with slip at the wall. It is the distance by which the wall would have to be displaced in the inviscid case to give the same total mass flow as the viscous case. The no-slip condition requires the flow velocity at the surface of a solid object be zero and the fluid temperature be equal to the temperature of the surface. The flow velocity will then increase rapidly within the boundary layer, governed by the boundary layer equations, below. The thermal boundary layer thickness is similarly the distance from the body at which the temperature is 99% of the temperature found from an inviscid solution. The ratio of the two thicknesses is governed by the Prandtl number. If the Prandtl number is 1, the two boundary layers are the same thickness. If the Prandtl number is greater than 1, the thermal boundary layer is thinner than the velocity boundary layer. If the Prandtl number is less than 1, which is the case for air at standard conditions, the thermal boundary layer is thicker than the velocity boundary layer.

In high-performance designs, such as gliders and commercial aircraft, much attention is paid to controlling the behaviour of the boundary layer to minimize drag. Two effects have to be considered. First, the boundary layer adds to the effective thickness of the body, through the displacement thickness, hence increasing the pressure drag. Secondly, the shear forces at the surface of the wing create skin friction drag.

At high Reynolds numbers, typical of full-sized aircraft, it is desirable to have a laminar boundary layer. This results in a lower skin friction due to the characteristic velocity profile of laminar flow. However, the boundary layer inevitably thickens and becomes less stable as the flow develops along the body, and eventually becomes turbulent, the process known as boundary layer transition. One way of dealing with this problem is to suck the boundary layer away through a porous surface. This can reduce drag, but is usually impractical due to its mechanical complexity and the power required to move the air and dispose of it. Natural laminar flow techniques push the boundary layer transition aft by reshaping the aerofoil or fuselage so that its thickest point is more aft and less thick. This reduces the velocities in the leading part and the same Reynolds number is achieved with a greater length.

At lower Reynolds numbers, such as those seen with model aircraft, it is relatively easy to maintain laminar flow. This gives low skin friction, which is desirable. However, the same velocity profile which gives the laminar boundary layer its low skin friction also causes it to be badly affected by adverse pressure gradients. As the pressure begins to recover over the rear part of the wing chord, a laminar boundary layer will tend to separate from the surface. Such flow separation causes a large increase in the pressure drag, since it greatly increases the effective size of the wing section. In these cases, it can be advantageous to deliberately trip the boundary layer into turbulence at a point prior to the location of laminar separation, using a turbulator.

The fuller velocity profile of the turbulent boundary layer allows it to sustain the adverse pressure gradient without separating. Thus, although the skin friction is increased, overall drag is decreased. This is the principle behind the dimpling on golf balls, as well as vortex generators on aircraft. Special wing sections have also been designed which tailor the pressure recovery so laminar separation is reduced or even eliminated. This represents an optimum compromise between the pressure drag from flow separation and skin friction from induced turbulence.

When using half-models in wind tunnels, a peniche is sometimes used to reduce or eliminate the effect of the boundary layer.

Naval Architecture

Many of the principles that apply to aircraft also apply to ships, submarines, and offshore platforms. For ships, unlike aircraft, one deals with incompressible flows, where change in water density is negligible (a pressure rises close to 1000kPa leads to a change of only 2–3 kg/m^3). This field of fluid dynamics is called hydrodynamics. A ship engineer designs for hydrodynamics first, and for strength only later. The boundary layer development, breakdown, and separation become critical because the high viscosity of water produces high shear stresses. Another consequence of high viscosity is the slip stream effect, in which the ship moves like a spear tearing through a sponge at high velocity.

Boundary Layer Equations

The deduction of the boundary layer equations was one of the most important advances in fluid dynamics (Anderson, 2005). Using an order of magnitude analysis, the well-known governing Navier–Stokes equations of viscous fluid flow can be greatly simplified within the boundary layer. Notably, the characteristic of the partial differential equations (PDE) becomes parabolic, rather than the elliptical form of the full Navier–Stokes equations. This greatly simplifies the solution of the equations. By making the boundary layer approximation, the flow is divided into an inviscid portion (which is easy to solve by a number of methods) and the boundary layer, which is governed by an easier to solve PDE. The continuity and Navier–Stokes equations for a two-dimensional steady incompressible flow in Cartesian coordinates are given by

$$\frac{\partial u}{\partial x}+\frac{\partial \upsilon}{\partial y}=0$$

$$u\frac{\partial u}{\partial x}+\upsilon\frac{\partial u}{\partial y}=-\frac{1}{\rho}\frac{\partial p}{\partial x}+\nu\left(\frac{\partial^2 u}{\partial x^2}+\frac{\partial^2 u}{\partial y^2}\right)$$

$$u\frac{\partial \upsilon}{\partial x}+\upsilon\frac{\partial \upsilon}{\partial y}=-\frac{1}{\rho}\frac{\partial p}{\partial y}+\nu\left(\frac{\partial^2 \upsilon}{\partial x^2}+\frac{\partial^2 \upsilon}{\partial y^2}\right)$$

where u and υ are the velocity components, ρ is the density, p is the pressure, and ν is the kinematic viscosity of the fluid at a point.

The approximation states that, for a sufficiently high Reynolds number the flow over a surface can be divided into an outer region of inviscid flow unaffected by viscosity (the majority of the flow), and a region close to the surface where viscosity is important (the boundary layer). Let u and υ be streamwise and transverse (wall normal) velocities respectively inside the boundary layer. Using scale analysis, it can be shown that the above equations of motion reduce within the boundary layer to become

$$\frac{\partial u}{\partial x}+\frac{\partial \upsilon}{\partial y}=0$$

$$u\frac{\partial u}{\partial x}+\upsilon\frac{\partial u}{\partial y}=-\frac{1}{\rho}\frac{\partial p}{\partial x}+\nu\frac{\partial^2 u}{\partial y^2}$$

and if the fluid is incompressible (as liquids are under standard conditions):

$$\frac{1}{\rho}\frac{\partial p}{\partial y}=0$$

The asymptotic analysis also shows that υ, the wall normal velocity, is small compared with u the streamwise velocity, and that variations in properties in the streamwise direction are generally much lower than those in the wall normal direction.

Since the static pressure p is independent of y, then pressure at the edge of the boundary layer is the pressure throughout the boundary layer at a given streamwise position. The external pressure may be obtained through an application of Bernoulli's equation. Let u_0 be the fluid velocity outside the boundary layer, where u and u_0 are both parallel. This gives upon substituting for p the following result

$$u\frac{\partial u}{\partial x}+\upsilon\frac{\partial u}{\partial y}=u_0\frac{\partial u_0}{\partial x}+\nu\frac{\partial^2 u}{\partial y^2}$$

with the boundary condition

$$\frac{\partial u}{\partial x}+\frac{\partial v}{\partial y}=0$$

For a flow in which the static pressure *p*also does not change in the direction of the flow then

$$\frac{\partial p}{\partial x}=0$$

so u_0 remains constant.

Therefore, the equation of motion simplifies to become

$$u\frac{\partial u}{\partial x}+\upsilon\frac{\partial u}{\partial y}=\nu\frac{\partial^2 u}{\partial y^2}$$

These approximations are used in a variety of practical flow problems of scientific and engineering interest. The above analysis is for any instantaneous laminar or turbulent boundary layer, but is used mainly in laminar flow studies since the mean flow is also the instantaneous flow because there are no velocity fluctuations present.

Turbulent Boundary Layers

The treatment of turbulent boundary layers is far more difficult due to the time-dependent variation of the flow properties. One of the most widely used techniques in which turbulent flows are tackled is to apply Reynolds decomposition. Here the instantaneous flow properties are decomposed into a mean and fluctuating component. Applying this technique to the boundary layer equations gives the full turbulent boundary layer equations not often given in literature:

$$\frac{\partial \overline{u}}{\partial x}+\frac{\partial \overline{v}}{\partial y}=0$$

$$\overline{u}\frac{\partial \overline{u}}{\partial x}+\overline{v}\frac{\partial \overline{u}}{\partial y}=-\frac{1}{\rho}\frac{\partial \overline{p}}{\partial x}+\nu\left(\frac{\partial^2 \overline{u}}{\partial x^2}+\frac{\partial^2 \overline{u}}{\partial y^2}\right)-\frac{\partial}{\partial y}(\overline{u'v'})-\frac{\partial}{\partial x}(\overline{u'^2})$$

$$\overline{u}\frac{\partial \overline{v}}{\partial x}+\overline{v}\frac{\partial \overline{v}}{\partial y}=-\frac{1}{\rho}\frac{\partial \overline{p}}{\partial y}+\nu\left(\frac{\partial^2 \overline{v}}{\partial x^2}+\frac{\partial^2 \overline{v}}{\partial y^2}\right)-\frac{\partial}{\partial x}(\overline{u'v'})-\frac{\partial}{\partial y}(\overline{v'^2})$$

Using the same order-of-magnitude analysis as for the instantaneous equations, these turbulent boundary layer equations generally reduce to become in their classical form:

$$\frac{\partial \overline{u}}{\partial x}+\frac{\partial \overline{v}}{\partial y}=0$$

$$\bar{u}\frac{\partial \bar{u}}{\partial x}+\bar{v}\frac{\partial \bar{u}}{\partial y}=-\frac{1}{\rho}\frac{\partial \bar{p}}{\partial x}+\nu\frac{\partial^2 \bar{u}}{\partial y^2}-\frac{\partial}{\partial y}(\overline{u'v'})$$

$$\frac{\partial \bar{p}}{\partial y}=0$$

The additional term $\overline{u'v'}$ in the turbulent boundary layer equations is known as the Reynolds shear stress and is unknown a priori. The solution of the turbulent boundary layer equations therefore necessitates the use of a turbulence model, which aims to express the Reynolds shear stress in terms of known flow variables or derivatives. The lack of accuracy and generality of such models is a major obstacle in the successful prediction of turbulent flow properties in modern fluid dynamics.

A laminar sub-layer exists in the turbulent zone; it occurs due to those fluid molecules which are still in the very proximity of the surface, where the shear stress is maximum and the velocity of fluid molecules is zero.

Heat and Mass Transfer

In 1928, the French engineer André Lévêque observed that convective heat transfer in a flowing fluid is affected only by the velocity values very close to the surface. For flows of large Prandtl number, the temperature/mass transition from surface to freestream temperature takes place across a very thin region close to the surface. Therefore, the most important fluid velocities are those inside this very thin region in which the change in velocity can be considered linear with normal distance from the surface. In this way, for

$$u(y)=u_0\left[1-\frac{(y-h)^2}{h^2}\right]=u_0\frac{y}{h}\left[2-\frac{y}{h}\right],$$

when $y\to 0$, then

$$u(y)\approx 2u_0\frac{y}{h}=\theta y,$$

where θ is the tangent of the Poiseuille parabola intersecting the wall. Although Lévêque's solution was specific to heat transfer into a Poiseuille flow, his insight helped lead other scientists to an exact solution of the thermal boundary-layer problem. Schuh observed that in a boundary-layer, u is again a linear function of y, but that in this case, the wall tangent is a function of x. He expressed this with a modified version of Lévêque's profile,

$$u(y) = \theta(x)y.$$

This results in a very good approximation, even for low P_r numbers, so that only liquid metals with P_r much less than 1 cannot be treated this way. In 1962, Kestin and Persen published a paper describing solutions for heat transfer when the thermal boundary layer is contained entirely within the momentum layer and for various wall temperature distributions.

For the problem of a flat plate with a temperature jump at $x = x_0$, they propose a substitution that reduces the parabolic thermal boundary-layer equation to an ordinary differential equation.

The solution to this equation, the temperature at any point in the fluid, can be expressed as an incomplete gamma function. Schlichting proposed an equivalent substitution that reduces the thermal boundary-layer equation to an ordinary differential equation whose solution is the same incomplete gamma function.

Convective Transfer Constants from Boundary Layer Analysis

Paul Richard Heinrich Blasius derived an exact solution to the above laminar boundary layer equations. The thickness of the boundary layer δ is a function of the Reynolds number for laminar flow.

$$\delta \approx \frac{5.0 * x}{\sqrt{Re}}$$

δ = the thickness of the boundary layer: the region of flow where the velocity is less than 99% of the far field velocity v_∞; x is position along the semi-infinite plate, and Re is the Reynolds Number given by $\rho v_\infty x / \mu$ (ρ = density and μ = dynamic viscosity).

The Blasius solution uses boundary conditions in a dimensionless form:

$$\frac{v_x - v_S}{v_\infty - v_S} = \frac{v_x}{v_\infty} = \frac{v_y}{v_\infty} = 0 \quad \text{at} \quad y = 0$$

$$\frac{v_x - v_S}{v_\infty - v_S} = \frac{v_x}{v_\infty} = 1 \quad \text{at} \quad y = \infty \text{ and } x = 0.$$

Note that in many cases, the no-slip boundary condition holds that v_S, the fluid velocity at the surface of the plate equals the velocity of the plate at all locations. If the plate is not moving, then $v_S = 0$. A much more complicated derivation is required if fluid slip is allowed.

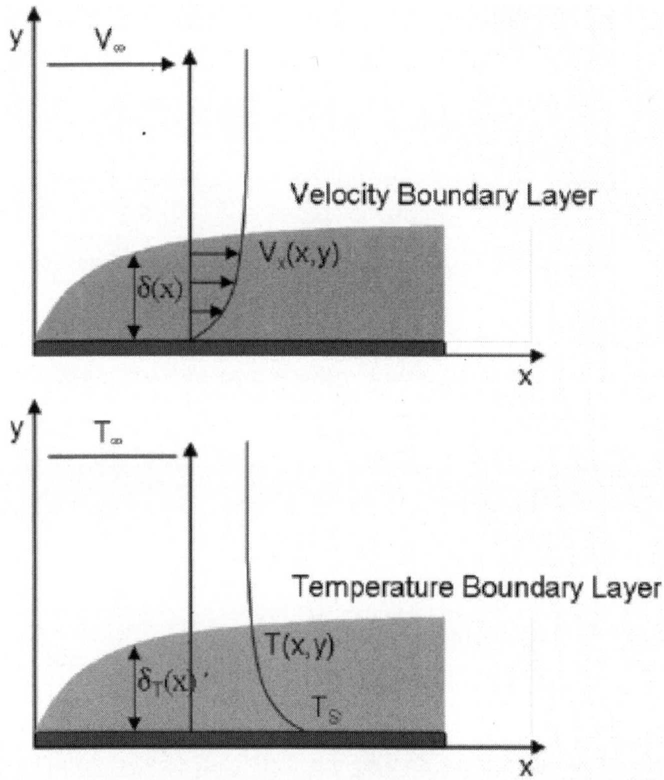

Figure: *Velocity Boundary Layer (Top,orange) and Temperature Boundary Layer (Bottom, green) share a functional form due to similarity in the Momentum/Energy Balances and boundary conditions.*

In fact, the Blasius solution for laminar velocity profile in the boundary layer above a semi-infinite plate can be easily extended to describe Thermal and Concentration boundary layers for heat and mass transfer respectively. Rather than the differential x-momentum balance (equation of motion), this uses a similarly derived Energy and Mass balance:

Energy: $$v_x \frac{\partial T}{\partial x} + v_y \frac{\partial T}{\partial y} = \frac{k}{\rho Cp} \frac{\partial^2 T}{\partial y^2}$$

Mass: $$v_x \frac{\partial c_A}{\partial x} + v_y \frac{\partial c_A}{\partial y} = D_{AB} \frac{\partial^2 c_A}{\partial y^2}$$

For the momentum balance, kinematic viscosity ν can be considered to be the *momentum diffusivity*. In the energy balance this is replaced by thermal diffusivity $\alpha = k / \rho C_P$, and by mass diffusivity D_{AB} in the mass balance. In thermal diffusivity of a substance, k is its thermal conductivity, ρ is its density and C_p is its heat capacity. Subscript AB denotes diffusivity of species A diffusing into species B.

Under the assumption that $\alpha = D_{AB} = \nu$, these equations become equivalent to the momentum balance. Thus, for Prandtl number $Pr = \nu / \alpha = 1$ and Schmidt number $Sc = \nu / D_{AB} = 1$ the Blasius solution applies directly.

Accordingly, this derivation uses a related form of the boundary conditions, replacing v with T or c_A (absolute temperature or concentration of species A). The subscript S denotes a surface condition.

$$\frac{v_x - v_S}{v_\infty - v_S} = \frac{T - T_S}{T_\infty - T_S} = \frac{c_A - c_{AS}}{c_{A\infty} - c_{AS}} = 0 \quad \text{at} \quad y = 0$$

$$\frac{v_x - v_S}{v_\infty - v_S} = \frac{T - T_S}{T_\infty - T_S} = \frac{c_A - c_{AS}}{c_{A\infty} - c_{AS}} = 1 \quad \text{at} \quad y = \infty \text{ and } x = 0$$

Using the streamline function Blasius obtained the following solution for the shear stress at the surface of the plate.

$$\tau_0 = \left(\frac{\partial v_x}{\partial y}\right)_{y=0} = 0.332 \frac{v_\infty}{x} Re^{1/2}$$

And via the boundary conditions, it is known that

$$\frac{v_x - v_S}{v_\infty - v_S} = \frac{T - T_S}{T_\infty - T_S} = \frac{c_A - c_{AS}}{c_{A\infty} - c_{AS}}$$

We are given the following relations for heat/mass flux out of the surface of the plate

$$\left(\frac{\partial T}{\partial y}\right)_{y=0} = 0.332 \frac{T_\infty - T_S}{x} Re^{1/2}$$

$$\left(\frac{\partial c_A}{\partial y}\right)_{y=0} = 0.332 \frac{c_{A\infty} - c_{AS}}{x} Re^{1/2}$$

So for $Pr = Sc = 1$

$$\delta = \delta_T = \delta_c = \frac{5.0 * x}{\sqrt{Re}}$$

Where δ_T, δ_c are the regions of flow where T and c_A are less than 99% of their far field values.

Because the Prandtl number of a particular fluid is not often unity, German engineer E. Polhausen who worked with Ludwig Prandtl attempted to empirically extend these equations to apply for $Pr \neq 1$. His results can be applied to Sc as well. He found that for Prandtl number greater than 0.6, the thermal boundary layer thickness was approximately given by:

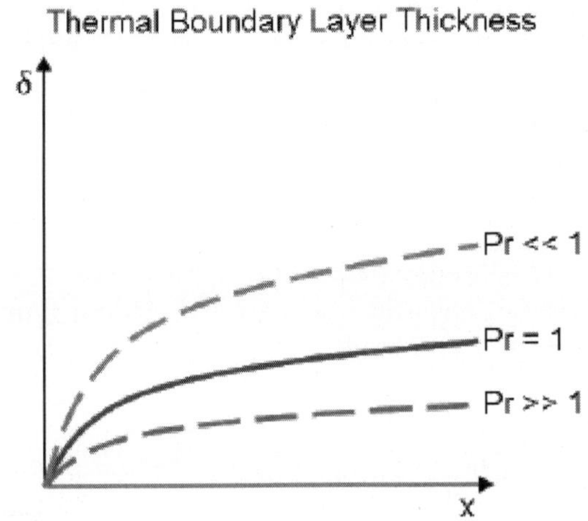

Figure: *Plot showing the relative thickness in the Thermal boundary layer versus the Velocity boundary layer (in red) for various Prandtl Numbers. For* $Pr = 1$, *the two are equal.*

$$\frac{\delta}{\delta_T} = Pr^{1/3} \quad \text{and therefore} \quad \frac{\delta}{\delta_c} = Sc^{1/3}$$

From this solution, it is possible to characterize the convective heat/mass transfer constants based on the region of boundary layer flow. Fourier's law of conduction and Newton's Law of Cooling are combined with the flux term derived above and the boundary layer thickness.

$$\frac{q}{A} = -k\left(\frac{\partial T}{\partial y}\right)_{y=0} = h_x(T_S - T_\infty)$$

$$h_x = 0.332\frac{k}{x}Re_x^{1/2}Pr^{1/3}$$

This gives the local convective constant h_x at one point on the semi-infinite plane. Integrating over the length of the plate gives an average

$$h_L = 0.664\frac{k}{x}Re_L^{1/2}Pr^{1/3}$$

Following the derivation with mass transfer terms (k = convective mass transfer constant, D_{AB} = diffusivity of species A into species B, $Sc = \nu / D_{AB}$), the following solutions are obtained:

$$k'_x = 0.332 \frac{D_{AB}}{x} Re_x^{1/2} Sc^{1/3}$$

$$k'_L = 0.664 \frac{D_{AB}}{x} Re_L^{1/2} Sc^{1/3}$$

These solutions apply for laminar flow with a Prandtl/Schmidt number greater than 0.6.

Boundary Layer Turbine

This effect was exploited in the Tesla turbine, patented by Nikola Tesla in 1913. It is referred to as a bladeless turbine because it uses the boundary layer effect and not a fluid impinging upon the blades as in a conventional turbine. Boundary layer turbines are also known as cohesion-type turbine, bladeless turbine, and Prandtl layer turbine (after Ludwig Prandtl).

Boundary-Layer Thickness

This page describes some parameters used to measure the properties of boundary layers. Consider a stationary body with a turbulent flow moving around it, like the semi-infinite flat plate with air flowing over the top of the plate. At the solid walls of the body the fluid satisfies a no-slip boundary condition and has zero velocity, but as you move away from the wall, the velocity of the flow asymptotically approaches the free stream mean velocity. Therefore it is impossible to define a clear point at which the boundary layer becomes the free stream. The parameters below overcome this and allow the boundary layer to be measured.

Motivations for Boundary Layers

As the Navier-Stokes equations that describe the motion of fluids are not solved in the general case, simplifying assumptions are used to describe approximate behaviour. For a fluid of low viscosity (compared to the geometrical configuration of the problem), the Reynolds number is large and the viscous term in the Navier-Stokes equation is small compared to the inertial term in much of the space. However, this is not true near the boundaries of the fluid and this description fails to capture some important behaviour in a lot of real systems, for instance the Kármán vortex street or the drag force on an airplane wing.

The boundary layer introduces a region in space (supposed to be a small fraction of the total space) near the solid obstacles where viscous behaviour dominates to solve the fluid equations in that region in the assumption of a small Reynolds number (i.e. viscosity dominates inertial effects). The boundary layer thickness is then some typical length scale that separates the two regions. It is derived from the solutions of fluid equations in the two domains by continuity of the equations : at the border between inside and outside the boundary layer, the speed of fluid must be continuous.

99% Boundary Layer Thickness

The boundary layer thickness, δ, is the distance across a boundary layer from the wall to a point where the flow velocity has essentially reached the 'free stream' velocity, u_0. This distance is defined normal to the wall, and the point where the flow velocity is essentially that of the free stream is customarily defined as the point where:

$$u(y) = 0.99u_o$$

For laminar boundary layers over a flat plate, the Blasius solution gives:

$$\delta \approx 4.91\sqrt{\frac{\nu x}{u_0}}$$

$$\delta \approx 4.91x / \sqrt{\mathrm{Re}_x}$$

For turbulent boundary layers over a flat plate, the boundary layer thickness is given by:

$$\delta \approx 0.382x / \mathrm{Re}_x^{1/5}$$

where

$\mathrm{Re}_x = \rho u_0 x / \mu$

δ is the overall thickness (or height) of the boundary layer

ν is the kinematic viscosity

x is the distance downstream from the start of the boundary layer

Re_x is the Reynolds Number

ρ is the density

u_0 is the freestream velocity

μ is the dynamic viscosity

The velocity thickness can also be referred to as the Soole ratio, although the gradient of the thickness over distance would be adversely proportional to that of velocity thickness.

Displacement Thickness

The displacement thickness, δ^* or δ_1 is the distance by which a surface would have to be moved in the direction perpendicular to its normal vector away from the reference plane in an inviscid fluid stream of velocity u_0 to give the same flow rate as occurs between the surface and the reference plane in a real fluid.

In practical aerodynamics, the displacement thickness essentially modifies the shape of a body immersed in a fluid to allow an inviscid solution. It is commonly used in aerodynamics to overcome the difficulty inherent in the fact that the fluid velocity in the boundary layer approaches asymptotically to the free stream value as distance from the wall increases at any given location.

The definition of the displacement thickness for compressible flow is based on mass flow rate:

$$\delta^* = \int_0^\infty \left(1 - \frac{\rho(y)u(y)}{\rho_0 u_0}\right) dy$$

The definition for incompressible flow can be based on volumetric flow rate, as the density is constant:

$$\delta^* = \int_0^\infty \left(1 - \frac{u(y)}{u_0}\right) dy$$

Where ρ_0 and u_0 are the density and velocity in the 'free stream' outside the boundary layer, and y is the coordinate normal to the wall.

For boundary layer calculations, the density and velocity at the edge of the boundary layer must be used, as there is no free stream. In the equations above, ρ_0 and u_0 are therefore replaced with ρ_e and u_e.

The displacement thickness is used to calculate the boundary layer's shape factor.

Momentum Thickness

The momentum thickness, θ or δ_2, is the distance by which a surface would have to be moved parallel to itself towards the reference plane in an inviscid fluid stream of velocity to give the same total momentum as exists between the surface and the reference plane in a real fluid. The definition of the momentum thickness for compressible flow is based on mass flow rate:

$$\theta = \int_0^\infty \frac{\rho(y)u(y)}{\rho_0 u_o}\left(1 - \frac{u(y)}{u_o}\right) dy$$

The definition for incompressible flow can be based on volumetric flow rate, as the density is constant:

$$\theta = \int_0^\infty \frac{u(y)}{u_o}\left(1 - \frac{u(y)}{u_o}\right)dy$$

Where ρ_0 and u_0 are the density and velocity in the 'free stream' outside the boundary layer, and y is the coordinate normal to the wall. For boundary layer calculations, the density and velocity at the edge of the boundary layer must be used, as there is no free stream. In the equations above, ρ_0 and u_0 are therefore replaced with ρ_e and u_e. For a flat plate at no angle of attack with a laminar boundary layer, the Blasius solution gives

$$\theta \approx 0.664\sqrt{\frac{\nu x}{u_o}}$$

The influence of fluid viscosity creates a wall shear stress, τ_w, which extracts energy from the mean flow. The boundary layer can be considered to possess a total momentum flux deficit, due to the frictional dissipation.

$$\rho\int_0^\infty u(y)(u_o - u(y))dy$$

Other length scales describing viscous boundary layers include the energy thickness, δ_3.

Shape Factor

A shape factor is used in boundary layer flow to determine the nature of the flow.

$$H = \frac{\delta^*}{\theta}$$

where H is the shape factor, δ^* is the displacement thickness and θ is the momentum thickness. The higher the value of H, the stronger the adverse pressure gradient. A high adverse pressure gradient can greatly reduce the Reynolds number at which transition into turbulence may occur. Conventionally, $H = 2.59$ (Blasius boundary layer) is typical of laminar flows, while H = 1.3 - 1.4 is typical of turbulent flows.

Boundary Layer Suction

Boundary layer suction is technique in which an air pump is used to extract the boundary layer at the wing or the inlet of an aircraft.

Improving the air flow can reduce drag. Improvements in fuel efficiency have been estimated as high as 30%.

The Boundary Layer

The air molecules at the surface of a wing are effectively stationary. If the flow is smooth, known as laminar flow, the velocity of the air increases steadily as measurements are taken further away from the surface. However the smooth flow is often disturbed by the boundary layer breaking away from the surface and creating a low pressure region immediately behind the airfoil. This low pressure region results in increased overall drag. Attempts have been made over the years to delay the onset of this flow separation by careful design and smooth surfaces.

Use of Suction

As flow separation results from the velocity deficit that is characteristic of boundary layers, suction attempts to remove the boundary layer from the surface before it can separate. The technology was first developed by Werner Pfenninger in the Second World War and has been researched almost continuously since. In the 1960s, NASA experimented with this concept with the Northrop X-21, a converted Douglas WB-66D. In the 1990s tests [1] were done by NASA with a F16XL.

Loek Boermans is researching the technology for use in gliders at the Technical University of Delft. However about 500 watts of power would be needed to drive the pumps and this would mean covering the glider with solar panels and would increase the cost greatly. There are also structural problems to be overcome before the ultimate glider could be manufactured.

Boundary Layer Control

Boundary layer control (BLC) refers to methods of controlling the behaviour of fluid flow boundary layers. This holds particular interest in aeronautical engineering because drag may be reduced whilst achieving high lift.

In the case of a freestream flow past a cylinder, a number of methods may be employed to control the boundary layer separation that occurs due to the adverse pressure gradient. Rotation of the cylinder can reduce or eliminate the boundary layer that is formed on the side which is moving in the same direction as the freestream. The side moving against the flow also exhibits only partial separation of the boundary layer. Suction applied through a slit in the cylinder near a separation point can also delay the onset of separation by removing fluid particles that have been slowed in the boundary layer. Alternatively, fluid can be blown from a faired slit such that the slowed fluid is accelerated and thus the point of separation is delayed.

In aeronautical engineering, boundary layer control refers to a number of methods of controlling the boundary layer of air on the main wing of an aircraft. In doing so, parasitic drag can be greatly reduced and performance likewise increased, while the usable angle of attack can be greatly increased, thereby dramatically improving lift at slow speeds. An aircraft with a boundary layer control system thus has greatly improved performance over a similar plane without such a system, often offering the otherwise contradictory features of STOL performance and high cruising speeds.

Much research was conducted to study the lift performance enhancement due to suction for aerofoils in the 1920s and 1930s at the Aerodynamische Versuchsanstalt in Goettingen. An example of an aircraft which uses BLC is the Japanese sea plane the ShinMaywa US-1. This large four-engined aircraft is used for anti-submarine warfare (ASW) and search and rescue (SAR). It is capable of STOL operation and very low air speeds, useful for both ASW and SAR.

Planetary Boundary Layer

The planetary boundary layer (PBL), also known as the atmospheric boundary layer (ABL), is the lowest part of the atmosphere and its behaviour is directly influenced by its contact with a planetary surface. On Earth it usually responds to changes in surface forcing in an hour or less. In this layer physical quantities such as flow velocity, temperature, moisture, etc., display rapid fluctuations (turbulence) and vertical mixing is strong. Above the PBL is the "free atmosphere" where the wind is approximately geostrophic (parallel to the isobars) while within the PBL the wind is affected by surface drag and turns across the isobars. The free atmosphere is usually nonturbulent, or only intermittently turbulent.

Cause of Surface Wind Gradient

Typically, due to aerodynamic drag, there is a wind gradient in the wind flow just a few hundred metres above the Earth's surface—the surface layer of the planetary boundary layer. Wind speed increases with increasing height above the ground, starting from zero due to the no-slip condition. Flow near the surface encounters obstacles that reduce the wind speed, and introduce random vertical and horizontal velocity components at right angles to the main direction of flow. This turbulence causes vertical mixing between the air moving horizontally at one level and the air at those levels immediately above and below it, which is important in dispersion of pollutants and in soil erosion.

The reduction in velocity near the surface is a function of surface roughness, so wind velocity profiles are quite different for different terrain types. Rough, irregular ground, and man-made obstructions on the ground can reduce the geostrophic wind speed by 40% to 50%. Over open water or ice, the reduction may be only 20% to 30%. These effects are taken into account when siting wind turbines. For engineering purposes, the wind gradient is modeled as a simple shear exhibiting a vertical velocity profile varying according to a power law with a constant exponential coefficient based on surface type. The height above ground where surface friction has a negligible effect on wind speed is called the "gradient height" and the wind speed above this height is assumed to be a constant called the "gradient wind speed". For example, typical values for the predicted gradient height are 457 m for large cities, 366 m for suburbs, 274 m for open terrain, and 213 m for open sea.

Although the power law exponent approximation is convenient, it has no theoretical basis. When the temperature profile is adiabatic, the wind speed should vary logarithmically with height. Measurements over open terrain in 1961 showed good agreement with the logarithmic fit up to 100 m or so (within the surface layer), with near constant average wind speed up through 1000 m.

The shearing of the wind is usually three-dimensional, that is, there is also a change in direction between the 'free' pressure-driven geostrophic wind and the wind close to the ground. This is related to the Ekman spiral effect. The cross-isobar angle of the diverted ageostrophic flow near the surface ranges from 10° over open water, to 30° over rough hilly terrain, and can increase to 40°-50° over land at night when the wind speed is very low.

After sundown the wind gradient near the surface increases, with the increasing stability. Atmospheric stability occurring at night with radiative cooling tends to contain turbulent eddies vertically, increasing the wind gradient. The magnitude of the wind gradient is largely influenced by the height of the convective boundary layer and this effect is even larger over the sea, where there is no diurnal variation of the height of the boundary layer as there is over land. In the convective boundary layer, strong mixing diminishes vertical wind gradient.

Constituent Layers

As Navier–Stokes equations suggest, the planetary boundary layer turbulence is produced in the layer with the largest velocity gradients that is at the very surface proximity. This layer – conventionally called a surface layer – constitutes about 10% of the total PBL depth. Above the surface layer the PBL turbulence gradually dissipates, losing its kinetic energy to friction as well as converting the kinetic to potential energy in a density stratified flow. The balance between the rate of the turbulent kinetic energy production and its dissipation determines the planetary boundary layer depth. The PBL depth varies broadly. At a given wind speed, e.g. 8 m/s, and so at a given rate of the turbulence production, a PBL in wintertime Arctic could be as shallow as 50 m, a nocturnal PBL in mid-latitudes could be typically 300 m in thickness, and a tropical PBL in the trade-wind zone could grow to its full theoretical depth of 2000 m.

In addition to the surface layer, the planetary boundary layer also comprises the PBL *core* (between 0.1 and 0.7 of the PBL depth) and the PBL top or *entrainment layer* or *capping inversion layer* (between 0.7 and 1 of the PBL depth). Four main external factors determine the PBL depth and its mean vertical structure:

1. the free atmosphere wind speed;
2. the surface heat (more exactly buoyancy) balance;
3. the free atmosphere density stratification;
4. the free atmosphere vertical wind shear or baroclinicity.

Principal Types

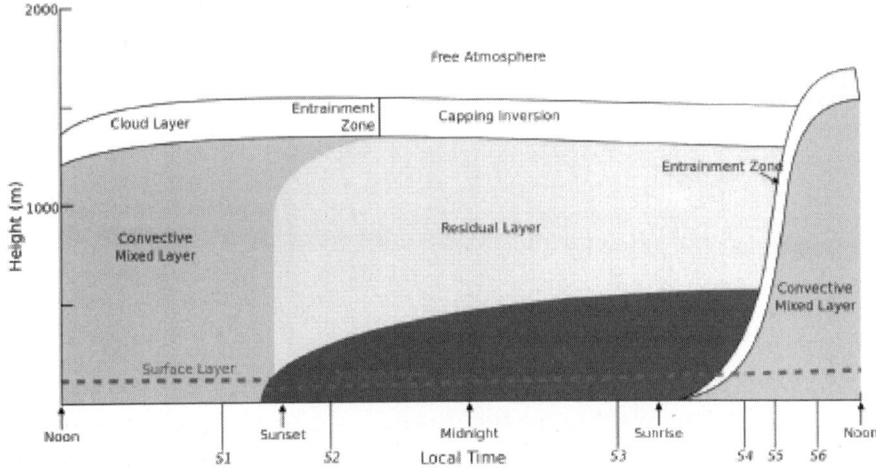

Convective Planetary Boundary Layer: (CBL) is the PBL where positive buoyancy flux at the surface creates a thermal instability and thus generates additional or even major turbulence. The CBL is typical in tropical and mid-latitudes during daytime. Solar heating assisted by the heat released from the water vapor condensation could create so strong convective turbulence that the CBL comprises the entire troposphere up to 10 km to 18 km (Intertropical convergence zone).

Stably Stratified Planetary Boundary Layer: (SBL) is the PBL where negative buoyancy flux at the surface damps the turbulence. The SBL is solely driven by the wind shear turbulence and hence the SBL cannot exist without the free atmosphere wind. The SBL is typical in nighttime at all locations and even in daytime in places where the Earth's surface is colder than the air above. The SBL plays a particularly important role in high latitudes where it is often prolonged (days to months), resulting in very cold air temperatures.

Physical laws and equations of motions, which govern the planetary boundary layer dynamics and microphysics, are strongly non-linear and considerably influenced by properties of the Earth's surface and evolution of the processes in the free atmosphere. To deal with this complicity, the whole array of turbulence modelling has been proposed. However, they are often not accurate enough to meet practical requests. Significant improvements are expected from application of a large eddy simulation technique to problems related to the PBL.

Perhaps the most important processes, which are critically dependent on the correct representation of the PBL in the atmospheric models (Atmospheric Model Intercomparison Project), are turbulent transport of moisture (evapotranspiration) and pollutants (air pollutants). Clouds in the boundary layer influence trade winds, the hydrological cycle, and energy exchange.

8

Earthquake Engineering

Earthquake engineering is the scientific field concerned with protecting society, the natural and the man-made environment from earthquakes by limiting the seismic risk to socioeconomically acceptable levels.

Traditionally, it has been narrowly defined as the study of the behaviour of structures and geo-structures subject to seismic loading, thus considered as a subset of both structural and geotechnical engineering. However, the tremendous costs experienced in recent earthquakes have led to an expansion of its scope to encompass disciplines from the wider field of civil engineering and from the social sciences, especially sociology, political sciences, economics and finance. The main objectives of earthquake engineering are:

- Foresee the potential consequences of strong earthquakes on urban areas and civil infrastructure.
- Design, construct and maintain structures to perform at earthquake exposure up to the expectations and in compliance with building codes.

A properly engineered structure does not necessarily have to be extremely strong or expensive. It has to be properly designed to withstand the seismic effects while sustaining an acceptable level of damage.

Seismic Loading

Seismic loading means application of an earthquake-generated excitation to a structure (or geo-structure). It happens at contact surfaces of a structure either with the ground, or with adjacent structures, or with gravity waves from tsunami.

Seismic Performance

Earthquake or seismic performance defines a structure's ability to sustain its due functions, such as its safety and serviceability, *at* and *after* a particular earthquake exposure. A structure is, normally, considered *safe* if it does not endanger the lives and well-being of those in or around it by partially or completely collapsing. A structure may be considered *serviceable* if it is able to fulfill its operational functions for which it was designed.

Basic concepts of the earthquake engineering, implemented in the major building codes, assume that a building should survive a rare, very severe earthquake by sustaining significant damage but without globally collapsing. On the other hand, it should remain operational for more frequent, but less severe seismic events.

Seismic Performance Assessment

Engineers need to know the quantified level of the actual or anticipated seismic performance associated with the direct damage to an individual building subject to a specified ground shaking. Such an assessment may be performed either experimentally or analytically.

Experimental Assessment

Experimental evaluations are expensive tests that are typically done by placing a (scaled) model of the structure on a shake-table that simulates the earth shaking and observing its behaviour. Such kinds of experiments were first performed more than a century ago. Still only recently has it become possible to perform 1:1 scale testing on full structures. Due to the costly nature of such tests, they tend to be used mainly for understanding the seismic behaviour of structures, validating modells and verifying analysis methods. Thus, once properly validated, computational modells and numerical procedures tend to carry the major burden for the seismic performance assessment of structures.

Analytical/Numerical Assessment

Seismic performance assessment or, simply, seismic structural analysis is a powerful tool of earthquake engineering which utilises detailed modelling of the structure together with methods of structural analysis to gain a better understanding of seismic performance of building and non-building structures. The technique as a formal concept is a relatively recent development. In general, seismic structural analysis is based on the methods of structural dynamics. For decades, the most prominent instrument of seismic analysis has been the earthquake response spectrum method which, also, contributed to the proposed building code's concept of today.

However, such methods are good only for linear elastic systems, being largely unable to model the structural behaviour when damage (i.e., non-linearity) appears. Numerical *step-by-step integration* proved to be a more effective method of analysis for multi-degree-of-freedom structural systems with significant non-linearity under a transient process of ground motion excitation.

Basically, numerical analysis is conducted in order to evaluate the seismic performance of buildings. Performance evaluations are generally carried out by using nonlinear static pushover analysis or nonlinear time-history analysis.

In such analyses, it is essential to achieve accurate nonlinear modelling of structural components such as beams, columns, beam-column joints, shear walls etc. Thus, experimental results play an important role in determining the modelling parametres of individual components, especially those that are subject to significant nonlinear deformations. The individual components are then assembled to create a full nonlinear model of the structure. Thus created modells are analysed to evaluate the performance of buildings.

The capabilities of the structural analysis software are a major consideration in the above process as they restrict the possible component modells, the analysis methods available and, most importantly, the numerical robustness.

The latter becomes a major consideration for structures that venture into the nonlinear range and approach global or local collapse as the numerical solution becomes increasingly unstable and thus difficult to reach. There are several commercially available Finite Element Analysis software's such as CSI-SAP2000 and CSI-PERFORM-3D which can be used for the seismic performance evaluation of buildings. Moreover, there is research-based finite element analysis platforms such as Open Sees, RUAUMOKO and the older DRAIN-2D/3D, several of which are now open source.

Research for Earthquake Engineering

Research for earthquake engineering means both field and analytical investigation or experimentation intended for discovery and scientific explanation of earthquake engineering related facts, revision of conventional concepts in the light of new findings, and practical application of the developed theories. The National Science

Foundation (NSF) is the main United States government agency that supports fundamental research and education in all fields of earthquake engineering. In particular, it focuses on experimental, analytical, and computational research on design and performance enhancement of structural systems.

Figure: *Shake-table testing of Friction Pendulum Bearings at EERC*

The Earthquake Engineering Research Institute (EERI) is a leader in dissemination of earthquake engineering research related information both in the U.S. and globally. A definitive list of earthquake engineering research related shaking tables around the world may be found in Experimental Facilities for Earthquake Engineering Simulation Worldwide. The most prominent of them is now E-Defence Shake Table in Japan. Earthquake engineering research activities worldwide are mostly associated with the following centres:

- Earthquake Engineering Research Institute (EERI)
- Earthquake Engineering Research Centre
- Pacific Earthquake Engineering Research Centre (PEER)
- John A. Blume Earthquake Engineering Centre
- Consortium of Universities for Research in Earthquake Engineering (CUREE)

- Multidisciplinary Centre for Earthquake Engineering Research (MCEER)
- George E. Brown, Jr. Network for Earthquake Engineering Simulation (NEES)
- USGS Earthquake Hazards Program
- Office of Earthquake Engineering at Caltrans
- Earthquake Engineering Research Centre of Iceland
- Earthquake Engineering New Zealand
- Canadian Research Centres and Research Groups on Earthquake Engineering
- Hyogo Earthquake Engineering Research Centre
- Laboratory for Earthquake Engineering of NTUA
- Earthquakes and Earthquake Engineering in The Library of Congress
- International Institute of Earthquake Engineering and Seismology
- National Centre for Research on Earthquake Engineering

Major U.S. Research Programs

The NSF Hazard Mitigation and Structural Engineering program (HMSE) supports research on new technologies for improving the behaviour and response of structural systems subject to earthquake hazards; fundamental research on safety and reliability of constructed systems; innovative developments in analysis and model based simulation of structural behaviour and response including soil-structure interaction; design concepts that improve structure performance and flexibility; and application of new control techniques for structural systems.

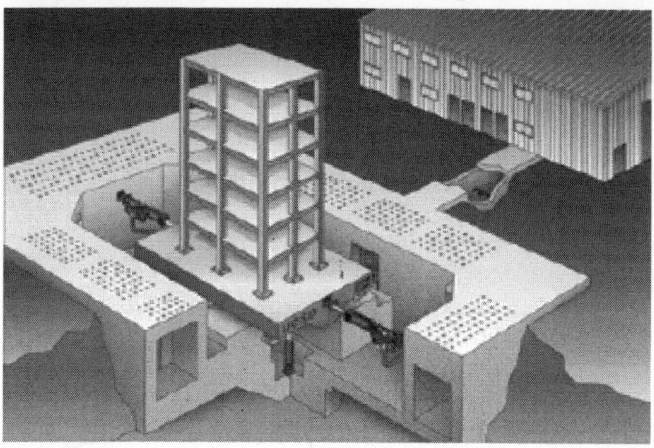

Figure : *Large High Performance Outdoor Shake Table, UCSD, NEES network.*

NSF also supports the George E. Brown, Jr. Network for Earthquake Engineering Simulation (NEES) that advances knowledge discovery and innovation for earthquakes and tsunami loss reduction of the nation's civil infrastructure, and new experimental simulation techniques and instrumentation. The NEES network features 14 geographically-distributed, shared-use laboratories that support several types of experimental work: geotechnical centrifuge research, shake-table tests, large-scale structural testing, tsunami wave basin experiments, and field site research. Participating universities include: Cornell University; Lehigh University; Oregon State University; Rensselaer Polytechnic Institute; University at Buffalo, SUNY; University of California, Berkeley; University of California, Davis; University of California, Los Angeles; University of California, San Diego; University of California, Santa Barbara; University of Illinois, Urbana-Champaign; University of Minnesota; University of Nevada, Reno; and the University of Texas, Austin.

The equipment sites (labs) and a central data repository are connected to the global earthquake engineering community via the NEES hub website. The NEES website is powered by HUBzero software developed at Purdue University for nanoHUB specifically to help the scientific community share resources and collaborate. The cyber infrastructure, connected via Internet 2, provides interactive simulation tools, a simulation tool development area, a curated central data repository, animated presentations, user support, telepresence, mechanism for uploading and sharing resources and statistics about users, and usage patterns. This cyber infrastructure allows researchers to: securely store, organize and share data within a standardised framework in a central location; remotely observe and participate in experiments through the use of synchronised real-time data and video; collaborate with colleagues to facilitate the planning, performance, analysis, and publication of research experiments; and conduct computational and hybrid simulations that may combine the results of multiple distributed experiments and link physical experiments with computer simulations to enable the investigation of overall system performance. These resources jointly provide the means for collaboration and discovery to improve the seismic design and performance of civil and mechanical infrastructure systems.

Earthquake Simulation

The very first earthquake simulations were performed by statically applying some *horizontal inertia forces* based on scaled peak ground accelerations to a mathematical model of a building. With the further development of computational technologies, static approaches began to give way to dynamic ones.

Dynamic experiments on building and non-building structures may be physical, like shake-table testing, or virtual ones. In both cases, to verify a structure's expected seismic performance, some researchers prefer to deal with so called "real time-histories" though the last cannot be "real" for a hypothetical earthquake specified by either a building code or by some particular research requirements. Therefore, there is a strong incentive to engage an earthquake simulation which is the seismic input that possesses only essential features of a real event. Sometimes, earthquake simulation is understood as a recreation of local effects of a strong earth shaking.

Structure Simulation

Theoretical or experimental evaluation of anticipated seismic performance mostly requires a structure simulation which is based on the concept of structural likeness or similarity. Similarity is some degree of analoguy or resemblance between two or more objects. The notion of similarity rests either on exact or approximate repetitions of patterns in the compared items.

Figure: *Concurrent experiments with two building models which are kinematically equivalent to a real prototype.*

In general, a building model is said to have similarity with the real object if the two share *geometric similarity*, *kinematic similarity* and *dynamic similarity*. The most vivid and effective type of similarity is the *kinematic* one. *Kinematic similarity* exists when the paths and velocities of moving particles of a model and its prototype are similar. The ultimate level of *kinematic similarity* is *kinematic equivalence* when, in the case of earthquake engineering, time-histories of each story lateral displacements of the model and its prototype would be the same.

Seismic Vibration Control

Seismic vibration control is a set of technical means aimed to mitigate seismic impacts in building and non-building structures. All seismic vibration control devices may be classified as *passive, active* or *hybrid* where:

- *passive control devices* have no feedback capability between them, structural elements and the ground;
- *active control devices* incorporate real-time recording instrumentation on the ground integrated with earthquake input processing equipment and actuators within the structure;
- *hybrid control devices* have combined features of active and passive control systems.

When ground seismic waves reach up and start to penetrate a base of a building, their energy flow density, due to reflections, reduces dramatically: usually, up to 90%. However, the remaining portions of the incident waves during a major earthquake still bear a huge devastating potential.

After the seismic waves enter a superstructure, there are a number of ways to control them in order to soothe their damaging effect and improve the building's seismic performance, for instance:

- to dissipate the wave energy inside a superstructure with properly engineered dampers;
- to disperse the wave energy between a wider range of frequencies;
- to absorb the resonant portions of the whole wave frequencies band with the help of so called *mass dampers*.

Devices of the last kind, abbreviated correspondingly as TMD for the tuned (*passive*), as AMD for the *active*, and as HMD for the *hybrid mass dampers*, have been studied and installed in high-rise buildings, predominantly in Japan, for a quarter of a century.

However, there is quite another approach: partial suppression of the seismic energy flow into the superstructure known as seismic or base isolation. For this, some pads are inserted into or under all major load-carrying elements in the base of the building which should substantially decouple a superstructure from its substructure resting on a shaking ground.

The first evidence of earthquake protection by using the principle of base isolation was discovered in Pasargadae, a city in ancient

Persia, now Iran: it goes back to 6th century BCE. Below, there are some samples of seismic vibration control technologies of today.

Dry-stone Walls Control

People of Inca civilization were masters of the polished '*dry-stone walls*', called ashlar, where blocks of stone were cut to fit together tightly without any mortar.

The Incas were among the best stonemasons the world has ever seen, and many junctions in their masonry were so perfect that even blades of grass could not fit between the stones. Peru is a highly seismic land, and for centuries the mortar-free construction proved to be apparently more earthquake-resistant than using mortar. The stones of the dry-stone walls built by the Incas could move slightly and resettle without the walls collapsing, a passive structural control technique employing both the principle of energy dissipation and that of suppressing resonant amplifications.

Lead Rubber Bearing

Lead Rubber Bearing or LRB is a type of base isolation employing a heavy damping. It was invented by Bill Robinson, a New Zealander.

Heavy damping mechanism incorporated in vibration control technologies and, particularly, in base isolation devices, is often considered a valuable source of suppressing vibrations thus enhancing a building's seismic performance. However, for the rather pliant systems such as base isolated structures, with a relatively low bearing stiffness but with a high damping, the so-called "damping force" may turn out the main pushing force at a strong earthquake. The video shows a Lead Rubber Bearing being tested at the UCSD Caltrans-SRMD facility. The bearing is made of rubber with a lead core. It was a uni-axial test in which the bearing was also under a full structure load. Many buildings and bridges, both in New Zealand and elsewhere, are protected with lead dampers and lead and rubber bearings. Te Papa Tongarewa, the national museum of New Zealand, and the New Zealand Parliament Buildings have been fitted with the bearings. Both are in Wellington, which sits on an active earthquake fault.

Tuned Mass Damper

Typically, the tuned mass dampers are huge concrete blocks mounted in skyscrapers or other structures and moved in opposition to the resonance frequency oscillations of the structures by means of some sort of spring mechanism.

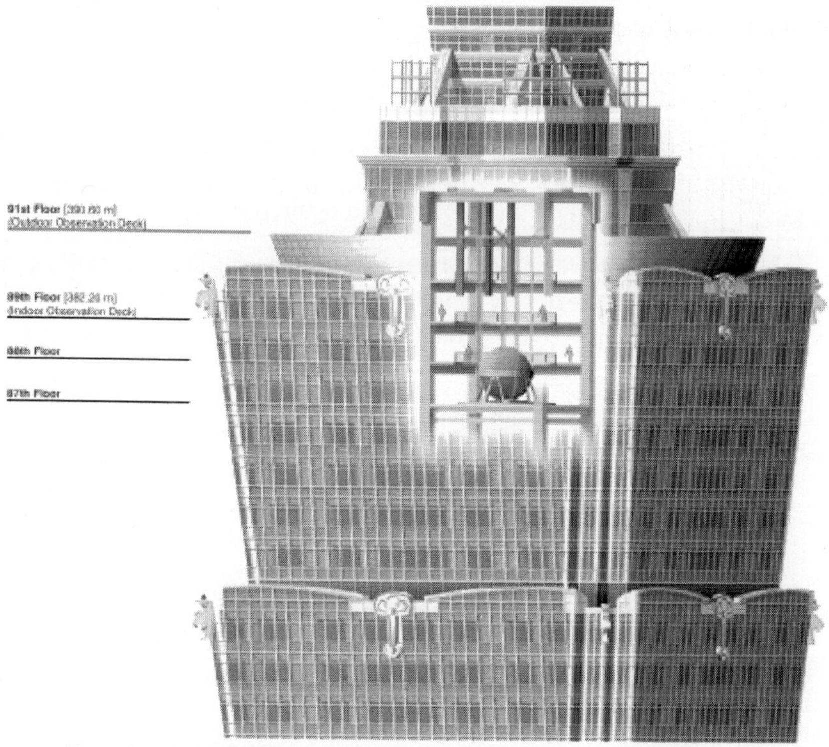

Figure: *Tuned mass damper in Taipei 101, the world's third tallest skyscraper*

Taipei 101 skyscraper needs to withstand typhoon winds and earthquake tremors common in its area of the Asia-Pacific.

For this purpose, a steel pendulum weighing 660 metric tons that serves as a tuned mass damper was designed and installed atop the structure.

Suspended from the 92nd to the 88th floor, the pendulums sways to decrease resonant amplifications of lateral displacements in the building caused by earthquakes and strong gusts.

Friction Pendulum Bearing

Friction Pendulum Bearing (FPB) is another name of Friction Pendulum System (FPS). It is based on three pillars:

- articulated friction slider;
- spherical concave sliding surface;
- enclosing cylinder for lateral displacement restraint.

Snapshot with the link to video clip of a shake-table testing of FPB system supporting a rigid building model is presented at the right.

Building Elevation Control

Building elevation control is a valuable source of vibration control of seismic loading. Pyramid-shaped skyscrapers continue to attract attention of architects and engineers because such structures promise a better stability against earthquakes and winds.

The elevation configuration can prevent buildings' resonant amplifications because a properly configured building disperses the shear wave energy between a wide range of frequencies.

Earthquake or wind quieting ability of the elevation configuration is provided by a specific pattern of multiple reflections and transmissions of vertically propagating shear waves, which are generated by breakdowns into homogeneity of story layers, and a taper.

Any abrupt changes of the propagating waves velocity result in a considerable dispersion of the wave energy between a wide ranges of frequencies thus preventing the resonant displacement amplifications in the building.

A tapered profile of a building is not a compulsory feature of this method of structural control. A similar resonance preventing effect can be also obtained by a proper *tapering* of other characteristics of a building structure, namely, its mass and stiffness.

As a result, the building elevation configuration techniques permit an architectural design that may be both attractive and functional.

Simple Roller Bearing

Simple roller bearing is a base isolation device which is intended for protection of various building and non-building structures against potentially damaging lateral impacts of strong earthquakes.

This metallic bearing support may be adapted, with certain precautions, as a seismic isolator to skyscrapers and buildings on soft ground. Recently, it has been employed under the name of *Metallic Roller Bearing* for a housing complex (17 stories) in Tokyo, Japan.

Springs-with-damper Base Isolator

Springs-with-damper base isolator installed under a three-story town-house, Santa Monica, California is shown on the photo taken prior to the 1994 Northridge earthquake exposure. It is a base isolation device conceptually similar to *Lead Rubber Bearing*.

Figure : *Springs-with-damper close-up.*

One of two three-story town-houses like this, which was well instrumented for recording of both vertical and horizontal accelerations on its floors and the ground, has survived a severe shaking during the Northridge earthquake and left valuable recorded information for further study.

Hysteretic Damper

Hysteretic damper is intended to provide better and more reliable seismic performance than that of a conventional structure at the expense of the seismic input energy dissipation. There are four major groups of hysteretic dampers used for the purpose, namely:

- Fluid viscous dampers (FVDs)
- Metallic yielding dampers (MYDs)
- Viscoelastic dampers (VEDs)
- Friction dampers (FDs).

Each group of dampers has specific characteristics, advantages and disadvantages for structural applications.

Seismic Design

Seismic design is based on authorised engineering procedures, principles and criteria meant to design or retrofit structures subject to earthquake exposure. Those criteria are only consistent with the contemporary state of the knowledge about earthquake engineering structures. Therefore, a building design which exactly follows seismic code regulations does not guarantee safety against collapse or serious damage.

The price of poor seismic design may be enormous. Nevertheless, seismic design has always been a trial and error process whether it was based on physical laws or on empirical knowledge of the structural performance of different shapes and materials.

To practice seismic design, seismic analysis or seismic evaluation of new and existing civil engineering projects, an engineer should, normally, pass examination on *Seismic Principles* which, in the State of California, include:

- Seismic Data and Seismic Design Criteria
- Seismic Characteristics of Engineered Systems
- Seismic Forces
- Seismic Analysis Procedures
- Seismic Detailing and Construction Quality Control.

To build up complex structural systems, seismic design largely uses the same relatively small number of basic structural elements (to say nothing of vibration control devices) as any non-seismic design project.

Normally, according to building codes, structures are designed to "withstand" the largest earthquake of a certain probability that is likely to occur at their location. This means the loss of life should be minimised by preventing collapse of the buildings.

Seismic design is carried out by understanding the possible failure modes of a structure and providing the structure with appropriate strength, stiffness, ductility, and configuration to ensure those modes cannot occur.

Seismic Design Requirements

Seismic design requirements depend on the type of the structure, locality of the project and its authorities which stipulate applicable seismic design codes and criteria. For instance, California Department of Transportation's requirements called *The Seismic Design Criteria* (SDC) and aimed at the design of new bridges in California incorporate an innovative seismic performance based approach.

***Figure** : Metsamor Nuclear Power Plant in Armenia was closed after the 1988 destructive earthquake.*

The most significant feature in the SDC design philosophy is a shift from a *force-based assessment* of seismic demand to a *displacement-based assessment* of demand and capacity. Thus, the newly adopted displacement approach is based on comparing the *elastic displacement* demand to the *inelastic displacement* capacity of the primary structural components while ensuring a minimum level of inelastic capacity at all potential plastic hinge locations.

In addition to the designed structure itself, seismic design requirements may include a *ground stabilisation* underneath the structure: sometimes, heavily shaken ground breaks up which leads to collapse of the structure sitting upon it. The following topics should be of primary concerns: liquefaction; dynamic lateral earth pressures on retaining walls; seismic slope stability; earthquake-induced settlement.

Nuclear facilities should not jeopardise their safety in case of earthquakes or other hostile external events. Therefore, their seismic design is based on criteria far more stringent than those applying to non-nuclear facilities. The Fukushima I nuclear accidents and damage to other nuclear facilities that followed the 2011 Tôhoku earthquake and tsunami have, however, drawn attention to ongoing concerns over Japanese nuclear seismic design standards and caused other many governments to re-evaluate their nuclear programs. Doubt has also been expressed over the seismic evaluation and design of certain other plants, including the Fessenheim Nuclear Power Plant in France.

Failure Modes

Failure mode is the manner by which an earthquake induced failure is observed. It, generally, describes the way the failure occurs. Though costly and time consuming, learning from each real earthquake failure remains a routine recipe for advancement in *seismic design* methods. Below, some typical modes of earthquake-generated failures are presented.

The lack of reinforcement coupled with poor mortar and inadequate roof-to-wall ties can result in substantial damage to a unreinforced masonry building. Severely cracked or leaning walls are some of the most common earthquake damage. Also hazardous is the damage that may occur between the walls and roof or floor diaphragms. Separation between the framing and the walls can jeopardize the vertical support of roof and floor systems.

Soft story effect. Absence of adequate shear walls on the ground level caused damage to this structure. A close examination of the image reveals that the rough board siding, once covered by a brick veneer, has been completely dismantled from the studwall. Only the rigidity of the floor above combined with the support on the two hidden sides by continuous walls, not penetrated with large doors as on the street sides, is preventing full collapse of the structure.

Soil liquefaction. In the cases where the soil consists of loose granular deposited materials with the tendency to develop excessive hydrostatic pore water pressure of sufficient magnitude and compact, liquefaction of those loose saturated deposits may result in non-uniform settlements and tilting of structures. This caused major damage to thousands of buildings in Niigata, Japan during the 1964 earthquake.

Landslide rock fall. A landslide is a geological phenomenon which includes a wide range of ground movement, including rock falls. Typically, the action of gravity is the primary driving force for a landslide to occur though in this case there was another contributing factor which affected the original slope stability: the landslide required an *earthquake trigger* before being released.

Pounding against adjacent building. This is a photograph of the collapsed five-story tower, St. Joseph's Seminary, Los Altos, California which resulted in one fatality. During Loma Prieta earthquake, the tower pounded against the independently vibrating adjacent building behind. A possibility of pounding depends on both buildings' lateral displacements which should be accurately estimated and accounted for.

At Northridge earthquake, the Kaiser Permanent concrete frame office building had joints completely shattered, revealing inadequate confinement steel, which resulted in the second story collapse. In the transverse direction, composite end shear walls, consisting of two wythes of brick and a layer of shotcrete that carried the lateral load, peeled apart because of inadequate through-ties and failed.

7 story reinforced concrete buildings on steep slope collapse due to the following :

- Improper construction site on a foothill.
- Poor detailing of the reinforcement (lack of concrete confinement in the columns and at the beam-column joints, inadequate splice length).
- Seismically weak soft story at the first floor.
- Long cantilevers with heavy dead load.

Sliding off foundations effect of a relatively rigid residential building structure during 1987 Whittier Narrows earthquake. The magnitude 5.9 earthquake pounded the Garvey West Apartment building in Monterey Park, California and shifted its superstructure about 10 inches to the east on its foundation. If a superstructure is not mounted on a base isolation system, its shifting on the basement should be prevented. Reinforced concrete column burst at Northridge earthquake due to insufficient shear reinforcement mode which allows main reinforcement to buckle outwards. The deck unseated at the hinge and failed in shear. As a result, the La Cienega-Venice underpass section of the 10 Freeway collapsed.

Loma Prieta earthquake: side view of reinforced concrete support-columns failure which trigged the upper deck collapse onto the lower deck of the two-level Cypress viaduct of Interstate Highway 880, Oakland, CA. Retaining wall failure at Loma Prieta earthquake in Santa Cruz Mountains area: prominent northwest trending extensional cracks up to 12 cm (4.7 in) wide in the concrete spillway to Austrian Dam, the north abutment.

Ground shaking triggered soil liquefaction in a subsurface layer of sand, producing differential lateral and vertical movement in an overlying carapace of unliquified sand and silt. This mode of ground failure, termed lateral spreading, is a principal cause of liquefaction-related earthquake damage.

Severely damaged building of Agriculture Development Bank of China after 2008 Sichuan earthquake: most of the beams and pier columns are sheared. Large diagonal cracks in masonry and veneer are due to in-plane loads while abrupt settlement of the right end of the building should be attributed to a landfill which may be hazardous even without any earthquake, see video footage at.

Twofold tsunami impact: sea waves hydraulic pressure and inundation. Thus, 2004 Indian Ocean earthquake of December 26, 2004, with the epicentre off the west coast of Sumatra, Indonesia, triggered a series of devastating tsunamis, killing more than 225,000 people in eleven countries by inundating surrounding coastal communities with huge waves up to 30 metres (100 feet) high.

Earthquake-Resistant construction

Earthquake construction means implementation of seismic design to enable building and non-building structures to live through the anticipated earthquake exposure up to the expectations and in compliance with the applicable building codes. Design and construction are intimately related. To achieve a good workmanship, detailing of the members and their connections should be, possibly, simple. As any construction in general, earthquake construction is a process that consists of the building, retrofitting or assembling of infrastructure given the construction materials available.

The destabilising action of an earthquake on constructions may be *direct* (seismic motion of the ground) or *indirect* (earthquake-induced landslides, soil liquefaction and waves of tsunami). A structure might have all the appearances of stability, yet offer nothing but danger when an earthquake occurs. The crucial fact is that, for safety, earthquake-resistant construction techniques are as important as quality control and using correct materials. *Earthquake contractor* should be registered in the state of the project location, bonded and insured. To minimize possible losses, construction process should be organised with keeping in mind that earthquake may strike any time prior to the end of construction.

Each construction project requires a qualified team of professionals who understand the basic features of seismic performance of different structures as well as construction management.

Adobe Structures

Around thirty percent of the world's population lives or works in earth-made construction. Adobe type of mud bricks is one of the oldest and most widely used building materials.

The use of adobe is very common in some of the world's most hazard-prone regions, traditionally across Latin America, Africa, Indian subcontinent and other parts of Asia, Middle East and Southern Europe. Adobe buildings are considered very vulnerable at strong quakes. However, multiple ways of seismic strengthening of new and existing adobe buildings are available, see, e.g.

Key factors for the improved seismic performance of adobe construction are:

- Quality of construction.
- Compact, box-type layout.
- Seismic reinforcement.

Limestone and Sandstone Structures

Limestone is very common in architecture, especially in North America and Europe. Many landmarks across the world, including the pyramids in Egypt, are made of limestone. Many medieval churches and castles in Europe are made of limestone and sandstone masonry. They are the long-lasting materials but their rather heavy weight is not beneficial for adequate seismic performance. Application of modern technology to seismic retrofitting can enhance the survivability of unreinforced masonry structures. As an example, from 1973 to 1989, the Salt Lake City and County Building in Utah was exhaustively renovated and repaired with an emphasis on preserving historical accuracy in appearance. This was done in concert with a seismic upgrade that placed the weak sandstone structure on base isolation foundation to better protect it from earthquake damage.

Timber Frame Structures

Timber framing dates back thousands of years, and has been used in many parts of the world during various periods such as ancient Japan, Europe and medieval England in localities where timber was in good supply and building stone and the skills to work it were not.

The use of timber framing in buildings provides their complete skeletal framing which offers some structural benefits as the timber frame, if properly engineered, lends itself to better *seismic survivability*.

Light-frame Structures

Figure : *A two-story wooden-frame for a residential building structure.*

Light-frame structures usually gain seismic resistance from rigid plywood shear walls and wood structural panel diaphragms. Special provisions for seismic load-resisting systems for all engineered wood structures requires consideration of diaphragm ratios, horizontal and vertical diaphragm shears, and connector/fastener values. In addition, collectors, or drag struts, to distribute shear along a diaphragm length are required.

Reinforced Masonry Structures

A construction system where steel reinforcement is embedded in the mortar joints of masonry or placed in holes and after filled with concrete or grout is called reinforced masonry. The devastating 1933 Long Beach earthquake revealed that masonry construction should be improved immediately. Then, the California State Code made the reinforced masonry mandatory. There are various practices and techniques to achieve reinforced masonry. The most common type is the reinforced hollow unit masonry. The effectiveness of both vertical and horizontal reinforcement strongly depends on the type and quality of the masonry, i.e. masonry units and mortar. To achieve a ductile behaviour of masonry, it is necessary that the shear strength of the wall is greater than the tensile strength of reinforcement to ensure a kind of bending failure.

Reinforced Concrete Structures

Reinforced concrete is concrete in which steel reinforcement bars (rubbers) or fibres have been incorporated to strengthen a material that would otherwise be brittle. It can be used to produce beams, columns, floors or bridges. Prestressed concrete is a kind of reinforced concrete used for overcoming concrete's natural weakness in tension.

It can be applied to beams, floors or bridges with a longer span than is practical with ordinary reinforced concrete. Prestressing tendons (generally of high tensile steel cable or rods) are used to provide a clamping load which produces a compressive stress that offsets the tensile stress that the concrete compression member would, otherwise, experience due to a bending load.

To prevent catastrophic collapse in response earth shaking (in the interest of life safety), a traditional reinforced concrete frame should have ductile joints. Depending upon the methods used and the imposed seismic forces, such buildings may be immediately usable, require extensive repair, or may have to be demolished.

Prestressed Structures

Prestressed structure is the one whose overall integrity, stability and security depend, primarily, on a *prestressing*. *Prestressing* means the intentional creation of permanent stresses in a structure for the purpose of improving its performance under various service conditions.

There are the following basic types of prestressing:

- Pre-compression (mostly, with the own weight of a structure)
- Pretensioning with high-strength embedded tendons
- Posttensioning with high-strength bonded or unbounded tendons.

Today, the concept of prestressed structure is widely engaged in design of buildings, underground structures, TV towers, power stations, floating storage and offshore facilities, nuclear reactor vessels, and numerous kinds of bridge systems. A beneficial idea of *prestressing* was, apparently, familiar to the ancient Rome architects; look, e.g., at the tall attic wall of Colosseum working as a stabilising device for the wall piers beneath.

Steel Structures

Steel structures are considered mostly earthquake resistant but their resistance should never be taken for granted. A great number of welded steel moment frame buildings, which looked earthquake-proof, surprisingly experienced brittle behaviour and were hazardously damaged in the 1994 Northridge earthquake. After that, the Federal Emergency Management Agency (FEMA) initiated development of repair techniques and new design approaches to minimize damage to steel moment frame buildings in future earthquakes.

For structural steel seismic design based on Load and Resistance Factor Design (LRFD) approach, it is very important to assess ability of a structure to develop and maintain its bearing resistance in the inelastic range. A measure of this ability is ductility, which may be observed in a *material itself*, in a *structural element*, or to a *whole structure.*

As a consequence of Northridge earthquake experience, all pre-qualified connection details and design methods contained in the building codes of that time have been rescinded. The new provisions stipulated that new designs be substantiated by testing or by use of test-verified calculations.

Prediction of Earthquake Losses

Earthquake loss estimation is usually defined as a *Damage Ratio* (DR) which is a ratio of the earthquake damage repair cost to the total value of a building. *Probable Maximum Loss* (PML) is a common term used for earthquake loss estimation, but it lacks a precise definition. In 1999, ASTM E2026 'Standard Guide for the Estimation of Building Damageability in Earthquakes' was produced in order to standardize the nomenclature for seismic loss estimation, as well as establish guidelines as to the review process and qualifications of the reviewer.

Earthquake loss estimations are also referred to as *Seismic Risk Assessments*. The risk assessment process generally involves determining the probability of various ground motions coupled with the vulnerability or damage of the building under those ground motions. The results are defined as a percent of building replacement value.

Environmental Engineering

Environmental engineering is the application of science and engineering principles to improve the natural environment (air, water, and/or land resources), to provide healthy water, air, and land for human habitation (house or home) and for other organisms, and to remediate polluted sites. It involves waste water management and air pollution control, recycling, waste disposal, radiation protection, industrial hygiene, environmental sustainability, and public health issues as well as a knowledge of environmental engineering law. It also includes studies on the environmental impact of proposed construction projects.

Environmental engineers conduct hazardous-waste management studies to evaluate the significance of such hazards, advise on treatment and containment, and develop regulations to prevent mishaps. Environmental engineers also design municipal water supply and industrial wastewater treatment systems as well as address local and worldwide environmental issues such as the effects of acid rain, global warming, ozone depletion, water pollution and air pollution from automobile exhausts and industrial sources. At many universities, Environmental Engineering programs follow either the Department of Civil Engineering or The Department of Chemical Engineering at Engineering faculties. Environmental "civil" engineers focus on hydrology, water resources management, bioremediation, and water treatment plant design. Environmental "chemical" engineers, on the other hand, focus on environmental chemistry, advanced air and water treatment technologies and separation processes.

Additionally, engineers are more frequently obtaining specialised training in law (J.D.) and are utilising their technical expertise in the practices of Environmental engineering law. About four percent of environmental engineers go on to obtain Board Certification in their speciality area(s) of environmental engineering (Board Certified Environmental Engineer or BCEE). Most jurisdictions also impose licensing and registration requirements.

Development of Environmental Engineering

Ever since people first recognised that their health and well-being were related to the quality of their environment, they have applied thoughtful principles to attempt to improve the quality of their environment. The ancient Harappan civilization utilised early sewers in some cities. The Romans constructed aqueducts to prevent drought and to create a clean, healthful water supply for the metropolis of Rome. In the 15th century, Bavaria created laws restricting the development and degradation of alpine country that constituted the region's water supply.

The field emerged as a separate environmental discipline during the middle third of the 20th century in response to widespread public concern about water and pollution and increasingly extensive environmental quality degradation. However, its roots extend back to early efforts in public health engineering. Modern environmental engineering began in London in the mid 19th century when Joseph Bazalgette designed the first major sewerage system that reduced the incidence of waterborne diseases such as cholera. The introduction of drinking water treatment and sewage treatment in industrialised countries reduced waterborne diseases from leading causes of death to rarities.

In many cases, as societies grew, actions that were intended to achieve benefits for those societies had longer-term impacts which reduced other environmental qualities. One example is the widespread application of DDT to control agricultural pests in the years following World War II. While the agricultural benefits were outstanding and crop yields increased dramatically, thus reducing world hunger substantially, and malaria was controlled better than it ever had been, numerous species were brought to the verge of extinction due to the impact of the DDT on their reproductive cycles. The story of DDT as vividly told in Rachel Carson's "Silent Spring" is considered to be the birth of the modern environmental movement and the development of the modern field of "environmental engineering." Conservation movements and laws restricting public actions that would harm the environment have been developed by various societies for millennia. Notable examples are the laws decreeing the construction of sewers in London and Paris in the 19th century and the creation of the U.S. national park system in the early 20th century.

Scope of Environmental Engineering

Briefly speaking, the main task of environmental engineers is to protect public health by protecting (from further degradation), preserving (the present condition of), and enhancing the environment. Environmental engineering is the application of science and engineering principles to the environment. Some consider environmental engineering to include the development of sustainable processes. There are several divisions of the field of environmental engineering.

Environmental Impact Assessment and Mitigation

In this division, engineers and scientists use a systemic identification and evaluation process to assess the potential impacts of a proposed project, plans, programs, policies, or legislative actions upon the physical-chemical, biological, cultural, and socioeconomic components on environmental conditions. They apply scientific and engineering principles to evaluate if there are likely to be any adverse impacts to water quality, air quality, habitat quality, flora and fauna, agricultural capacity, traffic impacts, social impacts, ecological impacts, noise impacts, visual (landscape) impacts, etc. If impacts are expected, they then develop mitigation measures to limit or prevent such impacts. An example of a mitigation measure would be the creation of wetlands in a nearby location to mitigate the filling in of wetlands necessary for a road development if it is not possible to reroute the road.

The practice of environmental assessment was intitiated on January 1, 1970, the effective date of the National Environmental Policy Act (NEPA) in the United States. Since that time, more than 100 developing and developed nations either have planned specific analoguous laws or have adopted procedure used elsewhere. NEPA is applicable to all federal agencies in the United States.

Water Supply and Treatment

Engineers and scientists work to secure water supplies for potable and agricultural use. They evaluate the water balance within a watershed and determine the available water supply, the water needed for various needs in that watershed, the seasonal cycles of water movement through the watershed and they develop systems to store, treat, and convey water for various uses. Water is treated to achieve water quality objectives for the end uses. In the case of potable water supply, water is treated to minimize the risk of infectious disease transmission, the risk of noninfectious illness, and to create a palatable water flavour. Water distribution systems are designed and built to provide adequate water pressure and flow rates to meet various end-user needs such as domestic use, fire suppression, and irrigation.

Wastewater Conveyance and Treatment

Most urban and many rural areas no longer discharge human waste directly to the land through outhouse, septic, and/or honey bucket systems, but rather deposit such waste into water and convey it from households via sewer systems. Engineers and scientists develop collection and treatment systems to carry this waste material away from where people live and produce the waste and discharge it into the environment. In developed countries, substantial resources are applied to the treatment and detoxification of this waste before it is discharged into a river, lake, or ocean system. Developing nations are striving to obtain the resources to develop such systems so that they can improve water quality in their surface waters and reduce the risk of waterborne infectious disease.

There are numerous wastewater treatment technologies. A wastewater treatment train can consist of a primary clarifier system to remove solid and floating materials, a secondary treatment system consisting of an aeration basin followed by flocculation and sedimentation or an activated sludge system and a secondary clarifier, a tertiary biological nitrogen removal system, and a final disinfection process. The aeration basin/activated sludge system removes organic material by growing bacteria (activated sludge). The secondary clarifier removes the activated sludge from the water. The tertiary system, although not always included due to costs, is becoming more prevalent to remove nitrogen and phosphorus and to disinfect the water before discharge to a surface water stream or ocean outfall.

Geotechnical Engineering

Geotechnical engineering is the branch of civil engineering concerned with the engineering behaviour of earth materials. Geotechnical engineering is important in civil engineering, but is also used by military, mining, petroleum, or any other engineering concerned with construction on or in the ground. Geotechnical engineering uses principles of soil mechanics and rock mechanics to investigate subsurface conditions and materials; determine the relevant physical/ mechanical and chemical properties of these materials; evaluate stability of natural slopes and man-made soil deposits; assess risks posed by site conditions; design earthworks and structure foundations; and monitor site conditions, earthwork and foundation construction.

A typical geotechnical engineering project begins with a review of project needs to define the required material properties. Then follows a site investigation of soil, rock, fault distribution and bedrock properties on and below an area of interest to determine their engineering properties including how they will interact with, on or in a proposed construction. Site investigations are needed to gain an understanding of the area in or on which the engineering will take place. Investigations can include the assessment of the risk to humans, property and the environment from natural hazards such as earthquakes, landslides, sinkholes, soil liquefaction, debris flows and rockfalls.

Ground Improvement refers to a technique that improves the engineering properties of the soil mass treated. Usually, the properties that are modified are shear strength, stiffness and permeability. Ground improvement has developed into a sophisticated tool to support foundations for a wide variety of structures. Properly applied, i.e. after giving due to consideration to the nature of the ground being improved and the type and sensitivity of the structures being built, ground improvement often reduces directs costs and saves time.

A geotechnical engineer then determines and designs the type of foundations, earthworks, and/or pavement subgrades required for the intended man-made structures to be built. Foundations are designed and constructed for structures of various sizes such as high-rise buildings, bridges, medium to large commercial buildings, and smaller structures where the soil conditions do not allow code-based design.

Foundations built for above-ground structures include shallow and deep foundations. Retaining structures include earth-filled dams and retaining walls. Earthworks include embankments, tunnels, dikes, levees, channels, reservoirs, deposition of hazardous waste and sanitary landfills. Geotechnical engineering is also related to coastal and ocean engineering. Coastal engineering can involve the design and construction of wharves, marinas, and jetties. Ocean engineering can involve foundation and anchor systems for offshore structures such as oil platforms. The fields of geotechnical engineering and engineering geology are closely related, and have large areas of overlap. However, the field of geotechnical engineering is a speciality of engineering, where the field of engineering geology is a speciality of geology.

History

Humans have historically used soil as a material for flood control, irrigation purposes, burial sites, building foundations, and as construction material for buildings. First activities were linked to irrigation and flood control, as demonstrated by traces of dykes, dams, and canals dating back to at least 2000 BCE that were found in ancient Egypt, ancient Mesopotamia and the Fertile Crescent, as well as around the early settlements of Mohenjo Daro and Harappa in the Indus valley. As the cities expanded, structures were erected supported by formalised foundations; Ancient Greeks notably constructed pad footings and strip-and-raft foundations. Until the 18 th century, however, no theoretical basis for soil design had been developed and the discipline was more of an art than a science, relying on past experience.

Several foundation-related engineering problems, such as the Leaning Tower of Pisa, prompted scientists to begin taking a more scientific-based approach to examining the subsurface. The earliest advances occurred in the development of earth pressure theories for the construction of retaining walls. Henri Gautier, a French Royal Engineer, recognised the "natural slope" of different soils in 1717, an idea later known as the soil's angle of repose. A rudimentary soil classification system was also developed based on a material's unit weight, which is no longer considered a good indication of soil type.

The application of the principles of mechanics to soils was documented as early as 1773 when Charles Coulomb (a physicist, engineer, and army Captain) developed improved methods to determine the earth pressures against military ramparts. Coulomb observed that, at failure, a distinct slip plane would form behind a sliding retaining wall and he suggested that the maximum shear stress on the slip plane, for design purposes, was the sum of the soil cohesion, c, and friction $\sigma\tan(\phi)$, where is the normal stress on the slip plane and is the friction angle of the soil. By combining Coulomb's theory with Christian Otto Mohr's 2D stress state, the theory became known as Mohr-Coulomb theory. Although it is now recognised that precise determination of cohesion is impossible because *c* is not a fundamental soil property, the Mohr-Coulomb theory is still used in practice today.

In the 19th century Henry Darcy developed what is now known as Darcy's Law describing the flow of fluids in porous media. Joseph Boussinesq (a mathematician and physicist) developed theories of stress distribution in elastic solids that proved useful for estimating stresses at depth in the ground; William Rankine, an engineer and physicist, developed an alternative to Coulomb's earth pressure theory. Albert Atterberg developed the clay consistency indices that are still used today for soil classification. Osborne Reynolds recognised in 1885 that shearing causes volumetric dilation of dense and contraction of loose granular materials.

Modern geotechnical engineering is said to have begun in 1925 with the publication of *Erdbaumechanik* by Karl Terzaghi (a civil engineer and geologist). Considered by many to be the father of modern soil mechanics and geotechnical engineering, Terzaghi developed the principle of effective stress, and demonstrated that the shear strength of soil is controlled by effective stress. Terzaghi also developed the framework for theories of bearing capacity of foundations, and the theory for prediction of the rate of settlement of clay layers due to consolidation. In his 1948 book, Donald Taylor recognised that interlocking and dilation of densely packed particles contributed to the peak strength of a soil. The interrelationships between volume change behaviour (dilation, contraction, and consolidation) and shearing behaviour were all connected via the theory of plasticity using critical state soil mechanics by Roscoe, Schofield, and Wroth with the publication of "On the Yielding of Soils" in 1958. Critical state soil mechanics is the basis for many contemporary advanced constitutive modells describing the behaviour of soil.

Practising Engineers

Geotechnical engineers are typically graduates of a four-year civil engineering program and often hold a masters degree. In the USA, geotechnical engineers are typically licensed and regulated as Professional Engineers (PEs) in most states; currently only California and Oregon have licensed geotechnical engineering specialities.

State governments will typically license engineers who have graduated from an ABET accredited school, passed the Fundamentals of Engineering examination, completed several years of work experience under the supervision of a licensed Professional Engineer, and passed the Professional Engineering examination.

Geotechnical Investigation

Geotechnical engineers perform geotechnical investigations to obtain information on the physical properties of soil and rock underlying (and sometimes adjacent to) a site to design earthworks and foundations for proposed structures, and for repair of distress to earthworks and structures caused by subsurface conditions. A geotechnical investigation will include surface exploration and subsurface exploration of a site. Sometimes, geophysical methods are used to obtain data about sites. Subsurface exploration usually involves in-situ testing (two common examples of in-situ tests are the standard penetration test and cone penetration test. In addition site investigation will often include subsurface sampling and laboratory testing of the soil samples retrieved. The digging of test pits and trenching (particularly for locating faults and slide planes) may also be used to learn about soil conditions at depth. Large diametre borings are rarely used due to safety concerns and expense, but are sometimes used to allow a geologist or engineer to be lowered into the borehole for direct visual and manual examination of the soil and rock stratigraphy.

A variety of soil samplers exist to meet the needs of different engineering projects. The standard penetration test (SPT), which uses a thick-walled split spoon sampler, is the most common way to collect disturbed samples. Piston samplers, employing a thin-walled tube, are most commonly used for the collection of less disturbed samples. More advanced methods, such as ground freezing and the Sherbrooke block sampler, are superior, but even more expensive.

Atterberg limits tests, water content measurements, and grain size analysis, for example, may be performed on disturbed samples obtained from thick walled soil samplers. Properties such as shear strength, stiffness hydraulic conductivity, and coefficient of consolidation may be significantly altered by sample disturbance. To measure these properties in the laboratory, high quality sampling is required. Common tests to measure the strength and stiffness include the triaxial shear and unconfined compression test.

Surface exploration can include geologic mapping, geophysical methods, and photogrammetry; or it can be as simple as an engineer walking around to observe the physical conditions at the site. Geologic mapping and interpretation of geomorphology is typically completed in consultation with a geologist or engineering geologist. Geophysical exploration is also sometimes used. Geophysical techniques used for subsurface exploration include measurement of seismic waves (pressure, shear, and Rayleigh waves), surface-wave methods and/or downhole methods, and electromagnetic surveys (magnetometre, resistivity, and ground-penetrating radar).

Foundations

A building's foundation transmits loads from buildings and other structures to the earth. Geotechnical engineers design foundations based on the load characteristics of the structure and the properties of the soils and/or bedrock at the site. In general, geotechnical engineers: 1) Estimate the magnitude and location of the loads to be supported; 2) Develop an investigation plan to explore the subsurface; 3) Determine necessary soil parametres through field and lab testing (e.g., consolidation test, triaxial shear test, vane shear test, standard penetration test); 4) Design the foundation in the safest and most economical manner.

The primary considerations for foundation support are bearing capacity, settlement, and ground movement beneath the foundations. Bearing capacity is the ability of the site soils to support the loads imposed by buildings or structures. Settlement occurs under all foundations in all soil conditions, though lightly loaded structures or rock sites may experience negligible settlements. For heavier structures or softer sites, both overall settlement relative to unbuilt areas or neighbouring buildings, and differential settlement under a single structure, can be concerns. Of particular concern is settlement which occurs over time, as immediate settlement can usually be compensated for during construction. Ground movement beneath a structure's foundations can occur due to shrinkage or swell of expansive soils due to climatic changes, frost expansion of soil, melting of permafrost, slope instability, or other causes. All these factors must be considered during design of foundations.

Many building codes specify basic foundation design parametres for simple conditions, frequently varying by jurisdiction, but such design techniques are normally limited to certain types of construction and certain types of sites, and are frequently very conservative.

In areas of shallow bedrock, most foundations may bear directly on bedrock; in other areas, the soil may provide sufficient strength for the support of structures. In areas of deeper bedrock with soft overlying soils, deep foundations are used to support structures directly on the bedrock; in areas where bedrock is not economically available, stiff "bearing layers" are used to support deep foundations instead.

Shallow Foundations

Shallow foundations are a type of foundation that transfers building load to the very near the surface, rather than to a subsurface layer. Shallow foundations typically have a depth to width ratio of less than 1.

Slab Foundations

A variant on spread footings is to have the entire structure bear on a single slab of concrete underlying the entire area of the structure. Slabs must be thick enough to provide sufficient rigidity to spread the bearing loads somewhat uniformly, and to minimize differential settlement across the foundation. In some cases, flexure is allowed and the building is constructed to tolerate small movements of the foundation instead. For small structures, like single-family houses, the slab may be less than 300 mm thick; for larger structures, the foundation slab may be several metres thick. Slab foundations can be either slab-on-grade foundations or embedded foundations, typically in buildings with basements. Slab-on-grade foundations must be designed to allow for potential ground movement due to changing soil conditions.

Deep Foundations

Deep foundations are used for structures or heavy loads when shallow foundations cannot provide adequate capacity, due to size and structural limitations. They may also be used to transfer building loads past weak or compressible soil layers. While shallow foundations rely solely on the bearing capacity of the soil beneath them, deep foundations can rely on end bearing resistance, frictional resistance along their length, or both in developing the required capacity. Geotechnical engineers use specialised tools, such as the cone penetration test, to estimate the amount of skin and end bearing resistance available in the subsurface.

There are many types of deep foundations including piles, drilled shafts, caissons, piers, and earth stabilised columns. Large buildings such as skyscrapers typically require deep foundations. For example, the Jin Mao Tower in China uses tubular steel piles about 1m (3.3 feet) driven to a depth of 83.5m (274 feet) to support its weight. In buildings that are constructed and found to undergo settlement, underpinning piles can be used to stabilise the existing building.

Lateral Earth Support Structures

A retaining wall is a structure that holds back earth. Retaining walls stabilize soil and rock from downslope movement or erosion and provide support for vertical or near-vertical grade changes. Cofferdams and bulkheads, structures to hold back water, are sometimes also considered retaining walls. The primary geotechnical concern in design and installation of retaining walls is that the retained material is attempting to move forward and downslope due to gravity. This creates soil pressure behind the wall, which can be analysed based on the angle of internal friction and the cohesive strength (c) of the material and the amount of allowable movement of the wall. This pressure is smallest at the top and increases toward the bottom in a manner similar to hydraulic pressure, and tends to push the wall forward and overturn it. Ground water behind the wall that is not dissipated by a drainage system causes an additional horizontal hydraulic pressure on the wall.

Gravity Walls

Gravity walls depend on the size and weight of the wall mass to resist pressures from behind. Gravity walls will often have a slight setback, or batter, to improve wall stability. For short, landscaping walls, gravity walls made from dry-stacked (mortarless) stone or segmental concrete units (masonry units) are commonly used.

Earlier in the 20th century, taller retaining walls were often gravity walls made from large masses of concrete or stone. Today, taller retaining walls are increasingly built as composite gravity walls such as: geosynthetic or steel-reinforced backfill soil with precast facing; gabions (stacked steel wire baskets filled with rocks), crib walls (cells built up log cabin style from precast concrete or timber and filled with soil or free draining gravel) or soil-nailed walls (soil reinforced in place with steel and concrete rods).

For reinforced-soil gravity walls, the soil reinforcement is placed in horizontal layers throughout the height of the wall. Commonly, the soil reinforcement is geogrid, a high-strength polymer mesh, that provide tensile strength to hold soil together. The wall face is often of precast, segmental concrete units that can tolerate some differential movement. The reinforced soil's mass, along with the facing, becomes the gravity wall. The reinforced mass must be built large enough to retain the pressures from the soil behind it. Gravity walls usually must be a minimum of 30 to 40 percent as deep (thick) as the height of the wall, and may have to be larger if there is a slope or surcharge on the wall.

Cantilever Walls

Prior to the introduction of modern reinforced-soil gravity walls, cantilevered walls were the most common type of taller retaining wall. Cantilevered walls are made from a relatively thin stem of steel-reinforced, cast-in-place concrete or mortared masonry (often in the shape of an inverted T). These walls cantilever loads (like a beam) to a large, structural footing; converting horizontal pressures from behind the wall to vertical pressures on the ground below. Sometimes cantilevered walls are buttressed on the front, or include a counterfort on the back, to improve their stability against high loads. Buttresses are short wing walls at right angles to the main trend of the wall. These walls require rigid concrete footings below seasonal frost depth. This type of wall uses much less material than a traditional gravity wall. Cantilever walls resist lateral pressures by friction at the base of the wall and/or passive earth pressure, the tendency of the soil to resist lateral movement.

Basements are a form of cantilever walls, but the forces on the basement walls are greater than on conventional walls because the basement wall is not free to move.

Excavation Shoring

Shoring of temporary excavations frequently requires a wall design which does not extend laterally beyond the wall, so shoring extends below the planned base of the excavation. Common methods of shoring are the use of sheet piles or soldier beams and lagging. Sheet piles are a form of driven piling using thin interlocking sheets of steel to obtain a continuous barrier in the ground, and are driven prior to excavation. Soldier beams are constructed of wide flange steel H sections spaced about 2–3 m apart, driven prior to excavation. As the excavation proceeds, horizontal timber or steel sheeting (lagging) is inserted behind the H pile flanges.

In some cases, the lateral support which can be provided by the shoring wall alone is insufficient to resist the planned lateral loads; in this case additional support is provided by walers or tie-backs. Walers are structural elements which connect across the excavation so that the loads from the soil on either side of the excavation are used to resist each other, or which transfer horizontal loads from the shoring wall to the base of the excavation. Tie-backs are steel tendons drilled into the face of the wall which extend beyond the soil which is applying pressure to the wall, to provide additional lateral resistance to the wall.

Bibliography

Armstrong, J. A.: *Forces Shaping the U.S. Academic Engineering Research Enterprise,* National Academy Press, Washington DC, 1995.

Armytage, W. H. G.: *A Social History of Engineering.* Cambridge: MIT Press, 1966.

Beaty H. Wayne: *Standard Handbook for Electrical Engineers.* New York, NY: McGraw-Hill, 2000.

Benjamin Henry: *The Engineering Drawings of Benjamin Henry Latrobe,* Yale University Press, NY, 1980.

Bennett, S.: *A History of Control Engineering: 1800-1930,* IEE Press, London, 1979.

Bhavikatti, S. S. : *A Textbook on Elements of Civil Engineering and Engineering Mechanics,* New Age International , Delhi, 2011.

Booker, P. J.: *A History of Engineering Drawing,* Chatto & Windus, London, 1963.

Calhoun, Daniel H. : *The American Civil Engineer: Origins and Conflict.* Cambridge: MIT Press, 1960.

Chen, W.F.: *Civil Engineering Handbook.* Boca Raton, FL: CRC Press, 1995.

David Fisher: *Rules of Thumb for Engineers and Scientists,* Gulf Pub, Houston, 1991.

Davison, C. S. C. B.: *Engineering Heritage,* Institute of Mechanical Engineers, London, 1963.

Domminghaus, H.: *Plastics for Engineers: Materials, Properties, Applications,* Hanser Publishers, Munich, 1993.

Dym, J.B.: *Product Design with Plastics: A Practical Manual,* Industrial Press, New York, 1983.

Edward H. Smith: *Mechanical Engineer's Reference Book,* Butterworth-Heinemann, Oxford, 1994.

Emerson, Howard P.: *Origins of Industrial Engineering: The Early Years of a Profession.* Institution of Industrial Engineers, 1988.

Fernandez, Martin : *Dictionary of Civil Engineering*, A P H, Delhi, 2011.

Fitchen, John: *Building Construction Before Mechanization.* MIT Press, 1986.

Florman, S.: *The Civilized Engineer*, St. Martin's Press, New York, 1979.

Groover, Mikell P.: *Fundamentals of Modern Manufacturing*, John Wiley & Sons, INC, New York, 2002.

Gupta, J.P. : *A T B of Engineering Mech. and Basic Civil Engineering*, S. Chand Publishing, Delhi, 2011.

Hasan, Syed Danish : *Civil Engineering Materials and Their Testing*, Narosa, Delhi, 2011.

Howard B., Cary,: *Modern Welding Technology*, Pearson Education, Chicago, New Jersey, 2005.

Joshi, Jagat Pati : *Harappan Architecture And Civil Engineering*, Rupa Pub, Delhi, 2008.

Kapoor, R : *Encyclopedia of Civil Engineering*, SBS Pub, Delhi, 2007.

MacCollum, David V.: *Construction Safety Planning.* New York, NY: Van Nostrand Reinhold, 1995.

Mendel, Otto: *Practical Piping Handbook.* Tulsa, OK: PennWell Books, 1981.

Nicholas P.: *Handbook of Chemical Engineering Calculations*, McGraw-Hill, New York, 2004.

Punmia, B.C.; Ashok Kr. Jain and Arun Kr. Jain: *Basic Civil Engineering*, Laxmi Publications, Delhi, 2003.

Richard Valentine: *Motor Control Electronics Handbook,* McGraw-Hill, New York, 1998.

Robert C. Rosaler: *Standard Handbook of Plant Engineering,* McGraw-Hill, New York, 1995.

Seely, Bruce E: *Building the American Highway System: Engineers as Policy Makers.* Philadelphia: Temple University Press, 1987.

Singh, Gurcharan : *Standard Handbook of Civil Engineering*, Standard Pub, Delhi, 2003.

Thomas Sixsmith & R. Hanselka: *Handbook of Thermoplastic Piping System Design.* Marcel Dekker Ltd, London, 1997.

Weman, Klas: *Welding Processes Handbook,* CRC Press, New York, 2003.

Index

❑❑❑